CliffsNotes®

Algebra II

CD includes 500+ practice problems in a Q&A flashcard format!

Practice Pack

By Mary Jane Sterling

Houghton Mifflin Harcourt
Boston   New York

Copyright © 2010 Houghton Mifflin Harcourt

Library of Congress Control Number: 2009933750

ISBN: 978-0-470-49597-1

Printed in the United States of America

DOO 10 9 8 7

4500540616

For information about permission to reproduce selections from this book, please write
Permissions, Houghton Mifflin Harcourt Publishing Company,
215 Park Avenue South, New York, New York 10003.

www.hmhco.com

Publisher's Acknowledgments

Editorial

Project Editor: Suzanne Snyder

Acquisitions Editor: Greg Tubach

Technical Editor: Ed Kohn

Editorial Assistant: Emily Hinkel

Composition

Project Coordinator: Kristie Rees

Indexer: BIM Indexing & Proofreading Services

Proofreader: Henry Lazarek

Wiley Publishing Composition Services

Table of Contents

Introduction

Mathematics is a language that has the advantage of being very precise, predictable, and dependable. The sum of two numbers is always the same. You can count on it. (Pardon the pun.) With the precision comes the necessity of knowing what the mathematical words mean. Words like *reduce, rationalize, simplify,* and others have meanings outside the world of mathematics and very specific meanings in algebra. Most of the mathematical words are defined within the particular discussions in the chapters, but the following pages contain material to get you started.

Terminology and Notation

Number Systems

Numbers are categorized by their characteristics and how they are formed.

Natural numbers are the numbers 1, 2, 3, ... Also called *counting numbers,* they are the first numbers a child learns.

Whole numbers are the numbers 0, 1, 2, 3, ... These are different from the *natural numbers* only by the addition of the number 0. Whole numbers contain no fractions or negatives.

Integers are the numbers . . . , $-3, -2, -1, 0, 1, 2, 3,$ . . . . The integers consist of all whole numbers and their opposites.

Rational numbers are numbers that can be written in the form $\frac{p}{q}$ where p and q are any integers, except that q cannot be 0. The decimal equivalent of any rational number is a value that either terminates or repeats (has the same pattern forever and doesn't end).

Irrational numbers are those numbers with a decimal value that never ends and never repeats. Some examples of *irrational numbers* are $\sqrt{2}, \sqrt{11}, \sqrt[3]{5},$ . . . . These are roots of numbers that are not perfect squares or cubes, and so on.

Real numbers are all of the *rational numbers* and *irrational numbers* combined.

Complex numbers are those numbers that can be written in the form $a + bi$ where a and b are real numbers and i is $\sqrt{-1}$. The bi term is called the "imaginary part" of the *complex number.*

Types of Expressions and Equations

Linear expressions and equations are those in which the terms are all first degree. The exponents are all 1s; the terms have one variable in them; and they are all raised to the power 1. Linear equations, in one variable, have only one solution. For example, $x + y = 1$, $y = 3x - 2$, and $x = 1$ are all linear expressions. Only the example $x = 1$ is a linear equation in one variable.

Quadratic expressions and equations are those with at least one term raised to the second degree and no term of higher degree. Quadratic equations can have two different solutions.

Cubic expressions and equations are those with at least one term raised to the third degree and no terms of higher degree. Cubic equations can have three different solutions.

Nth degree expressions and equations are those with at least one term raised to the *n*th degree and no terms of higher degree. The equations can have as many solutions as the degree number.

Notation

Ordered pairs, ordered triples, and so on can be used to express solutions for systems of equations. Instead of writing $x = 2$, $y = 3$ or $a = 1$, $b = -1$, $c = 7$, the ordered pair $(2,3)$ or the ordered triple $(1, -1, 7)$ can be used. The ordered pair and triple are much simpler to write, but there has to be an agreement between the writer and the reader that the *order* in which the values are written is set—it cannot be changed, or you wouldn't know which number represents which variable.

Interval notation is an alternative to *inequality notation.* Instead of writing the inequality notation $-1 < x \leq 4$, you can write the interval notation $(-1,4]$. The parenthesis is used to show $<$ or $>$, meaning that the number is not included. The bracket is used to show $\leq$ or $\geq$, meaning that the number is included. To write $x > 5$, the interval notation is $(5,\infty)$, showing that there is no end to how large x can be. A parenthesis is used with ∞.

Absolute value notation consists of two vertical bars around a number or expression. $|4|$ or $|-3|$ means to find the distance from the value inside the bars to zero. $|4| = 4, |-3| = 3$. The definition of the absolute value operation is

$$|a| = \begin{cases} a, \text{ if } a \geq 0 \\ -a, \text{ if } a < 0 \end{cases}$$

An important equivalence is $\sqrt{a^2} = |a|$.

Note: The negative of a negative is a positive.

Order of Operations

When two or more numbers and/or variables are being combined with operations, there is a precise order in which these operations must be performed. This is called the *order of operations*. It is universally accepted. Any mathematician, anywhere in the world, will know what another mathematician means because they both use the same rules for combining the values.

Order

1. Perform all powers and roots, moving from left to right.

2. Perform all multiplication and division, moving from left to right.

3. Perform all addition and subtraction, moving from left to right.

For example, to simplify $3 \cdot 4 - 5 \cdot 2^3 + \frac{12}{3} + \sqrt{49}$, first raise the 2 to the third power and find the square root of 49: $3 \cdot 4 - 5 \cdot 8 + \frac{12}{3} + 7$. Next, multiply the 3 and 4, multiply the 5 and 8, and divide the 12 and 3: $12 - 40 + 4 + 7$. Lastly, subtract the 40 from the 12 and add the 4 and 7: $12 - 40 + 4 + 7 = -28 + 4 + 7 = -24 + 7 = -17$.

Grouping Symbols

The order of operations is *interrupted* or overridden if there are grouping symbols. The grouping symbols are parentheses ( ), brackets [ ], braces { }, radicals $\sqrt{}$, fraction lines $\frac{a}{b}$, and absolute value bars | |. When these grouping symbols occur, the operations within the grouping symbols take precedence—they have to be done first before the result can be combined with the other values.

Graphing and the Coordinate Plane

Quadrants

The coordinate plane is divided by the two axes into four quadrants, labeled counterclockwise, beginning with the upper right quadrant. The first quadrant (upper right) has points with both the x- and y-coordinates positive in value. The second quadrant (upper left) has points with the x-value negative and the y-value positive. The third quadrant (lower left) has points with both the x- and y-values negative. And the fourth quadrant (lower right) has points with the x-coordinate positive and the y-coordinate negative.

The vertical axis is usually the y-axis, and the horizontal axis is usually the x-axis. They meet at the origin, which has the coordinates (0,0).

Formulas and Forms

Formulas involving the points on the coordinate plane are as follows:

Distance between two points, (x_1, y_1) and (x_2, y_2): $d = \sqrt{\left(x_2 - x_1\right)^2 + \left(y_2 - y_1\right)^2}$

Midpoint of the segment between two points, (x_1, y_1) and (x_2, y_2): $M = \left(\dfrac{x_2 + x_1}{2}, \dfrac{y_2 + y_1}{2} \right)$

Slope of the line through two points, (x_1, y_1) and (x_2, y_2): $m = \dfrac{y_2 - y_1}{x_2 - x_1}$

Forms for the equation of a line are as follows:

Slope-intercept form: $y = mx + b$ where m is the slope of the line and b is the y-intercept.
Standard form: $ax + by = c$ where a, b, and c are constants.

Simplifying Fractions

Complex Fractions

A complex fraction is one that has a fraction in the numerator, the denominator, or both. Two effective methods are used to simplify them. One is to multiply the numerator of the fraction by the reciprocal of the denominator. The other method involves multiplying the numerator and denominator by the same value; this method is used when there happens to be a nice, convenient value that can be used in the multiplication.

Multiplying the Numerator by the Reciprocal of the Denominator

$$\frac{\frac{2}{3}}{\frac{4}{7}} = \frac{2}{3} \cdot \frac{7}{4} = \frac{\cancel{2}^{1}}{3} \cdot \frac{7}{\cancel{4}_{2}} = \frac{7}{6}$$

In this case, the result is a single, improper fraction.

$$\frac{\frac{a}{b}+1}{2+\frac{c}{d}}=\frac{\frac{a}{b}+\frac{b}{b}}{\frac{2d}{d}+\frac{c}{d}}=\frac{\frac{a+b}{b}}{\frac{2d+c}{d}}=\frac{a+b}{b}\cdot\frac{d}{2d+c}=\frac{(a+b)d}{b(2d+c)}$$

In this case, the numerator and denominator had to be simplified into single fractions before the multiplication could be performed.

Multiplying Numerator and Denominator by the Same Value

$$\frac{\frac{x+1}{5}}{\frac{2x-1}{10}}=\frac{\frac{x+1}{5}}{\frac{2x-1}{10}}\cdot\frac{10}{10}=\frac{\frac{x+1}{5}}{\frac{2x-1}{10}}\cdot\frac{\overset{2}{\cancel{10}}}{1}=\frac{(x+1)2}{2x-1}=\frac{2x+2}{2x-1}$$

The denominators of the fractions in both the numerator and denominator had numbers that divided 10 evenly. This provided a convenient number by which to multiply.

Rationalizing Fractions

Fractional values usually are written without a radical in the denominator. If a single term is in the denominator that has a radical, then multiply by a radical to simplify the fraction. If more than one term exists, with one or more of them radicals, then multiply by the conjugate of the denominator.

One Term

$$\frac{3}{\sqrt6}=\frac{3}{\sqrt6}\cdot\frac{\sqrt6}{\sqrt6}=\frac{3\sqrt6}{6}=\frac{\sqrt6}{2}$$

More Than One Term

$$\frac{2}{\sqrt3+1}=\frac{2}{\sqrt3+1}\cdot\frac{\sqrt3-1}{\sqrt3-1}=\frac{2(\sqrt3-1)}{3-1}=\frac{\cancel{2}(\sqrt3-1)}{\cancel{2}}=\sqrt3-1$$

$$\frac{18}{\sqrt6-\sqrt2}=\frac{18}{\sqrt6-\sqrt2}\cdot\frac{\sqrt6+\sqrt2}{\sqrt6+\sqrt2}=\frac{18(\sqrt6+\sqrt2)}{6-2}=\frac{9}{2}(\sqrt6+\sqrt2)$$

Common Errors

The following is a list of some of the more common errors made in algebra.

Error 1: $\sqrt{a^2+b^2}\neq a+b$

Consider $\sqrt{3^2+4^2}$. This is not equal to 3 + 4 or 7. By the order of operations, you simplify the expression under the radical first: $\sqrt{3^2+4^2}=\sqrt{9+16}=\sqrt{25}=5$.

Error 2: $(a + b)^2 \neq a^2 + b^2$ and $(a + b)^3 \neq a^3 + b^3$

$(a + b)^2 = a^2 + 2ab + b^2$ and $(a + b)^3 = a^3 + 3a^2b + 3ab^2 + b^3$. Look at $(1 + 2)^2$ and $(1 + 2)^3$. Again, by the order of operation, you simplify in the parentheses first: $(1 + 2)^2 = 3^2 = 9$ and $(1 + 2)^3 = 3^3 = 27$. Doing these the wrong way would result in answers of $1^2 + 2^2 = 1 + 4 = 5$ and $1^3 + 2^3 = 1 + 8 = 9$. See Chapter 1 for more on the powers of the binomial $(a + b)^n$.

Error 3: $-a^2 \neq a^2$

$-3^2 = -9$ and $(-3)^2 = 9$. The big difference here is the parentheses. In the first equation, the order of operations dictates that you raise to the power first and then take the negative. In the second equation, the parentheses mean that the number -3 is multiplying times itself.

Error 4: $\dfrac{-6 \pm \sqrt{5}}{2} \neq -3 \pm \sqrt{5}$

The original fraction shows that both the -6 and the $\sqrt{5}$ are divided by the 2. You can't reduce the fraction by dividing just one of the terms in the numerator by the denominator. The correct way to reduce this is $\dfrac{-6 \pm \sqrt{5}}{2} = \dfrac{-6}{2} \pm \dfrac{\sqrt{5}}{2} = -3 \pm \dfrac{\sqrt{5}}{2}$.

Error 5: $-2(x + y - 3) \neq -2x + y - 3$

A common error when distributing is to forget to multiply every term in the parentheses by the multiplier. The correct answer here is $-2(x + y - 3) = -2x - 2y + 6$.

Error 6: $\dfrac{6}{2 + 3} \neq \dfrac{6}{2} + \dfrac{6}{3} = 3 + 2 = 5$

You can't break up the fraction like this. If, however, the sum is in the numerator, then a breakup is allowed: $\dfrac{2 + 3}{6} = \dfrac{2}{6} + \dfrac{3}{6}$.

Pretest

Pretest Questions

In problems 1 through 5, simplify the expression.

1. $\dfrac{x^2 x^5}{\left(x^4\right)^3}$

   A. x^{-2}

   B. x^{-5}

   C. 1

2. $\dfrac{a^{-2} b^3 c}{ab^{-2} c^{-1}}$

   A. $\dfrac{b^5 c^2}{a^3}$

   B. $\dfrac{b}{ac^2}$

   C. $\dfrac{b}{a^2 c^2}$

3. $\left(\dfrac{ax^2}{y}\right)^2 \left(\dfrac{y^2}{a^2 x}\right)^3$

   A. $\left(\dfrac{ax^2 y^2}{a^2 xy}\right)^5$

   B. $\dfrac{x^6 y^6}{a^6}$

   C. $\dfrac{xy^4}{a^4}$

4. $2a^2 + 3a - 4a^2 + 2a + 5$

   A. $-2a^2 + 5a + 5$

   B. $5a^2 - 2a + 5$

   C. $2a^2 + 5a + 5$

5. $4x^2 + 2y^2 + 3x + 4y^2 - 5x$

   A. $4x^2 + 6y^2 - 2x$

   B. $-2x^2 + 6y^2$

   C. $7x^2 + 8y^2 - 5x$

In problems 6 through 11, find the product and simplify the answer.

6. $(8a - 3)(a + 1)$

   A. $8a^2 - 11a - 3$

   B. $8a^2 + 5a - 3$

   C. $8a^2 - 5a - 3$

7. $(6x^2 - y)(4x^2 - 5y)$

   A. $24x^4 - 15x^2y + 5y^2$

   B. $24x^4 - 31x^2y + 5y^2$

   C. $24x^4 - 34x^2y + 5y^2$

8. $(mn - 1)(mn + 1)$

   A. $mn - 1$

   B. $mn^2 - 1$

   C. $m^2n^2 - 1$

9. $(2 - y)^2$

   A. $4 - y^2$

   B. $4 - 4y + y^2$

   C. $4 - 2y - y^2$

10. $(x + y)^3$

    A. $x^3 + y^3$

    B. $x^3 + x^2y + xy^2 + y^3$

    C. $x^3 + 3x^2y + 3xy^2 + y^3$

11. $(3 - p)(9 + 3p + p^2)$

    A. $27 - p^3$

    B. $27 - 6p + 9p^2 - p^3$

    C. $27 - 3p + 3p^2 - p^3$

In problems 12 and 13, solve the division problem and write any remainder as a fraction.

12. $(6x^4 + 3x^3 - 2x^2 + 5x - 3) \div (x + 1)$

    A. $6x^3 - 3x^2 + x + 4 + \dfrac{-7}{x + 1}$

    B. $6x^3 + 3x^2 - 2x + 5 + \dfrac{3}{x + 1}$

    C. $6x^3 + 2x^2 - x + 4 + \dfrac{-2}{x + 1}$

13. $(3y^3 + 5y^2 + 10y - 4) \div (3y - 1)$
    A. $y^2 + 2y + 10 + \dfrac{-4}{3y - 1}$
    B. $y^2 + 2y + 4$
    C. $y^2 - 2y + 4 + \dfrac{1}{3y - 1}$

In problems 14 through 20, factor the expression completely.

14. $6x^2y + 4xy^3 + 10x^4y^4$
    A. $2xy(3x + 2y^2 + 5x^3y^3)$
    B. $2xy(3x + 4y^2 + 10x^2y^3)$
    C. $2xy(3x + 2y + 5xy)$

15. $25y^2 - 9$
    A. $(25y - 3)(y + 3)$
    B. $(5y - 3)(5y - 3)$
    C. $(5y - 3)(5y + 3)$

16. $27m^3 - 8$
    A. $(3m - 2)^3$
    B. $(3m - 2)(9m^2 + 6m + 4)$
    C. $(3m - 2)(9m + 4)$

17. $10y^2 - y - 3$
    A. $(5y - 3)(2y + 1)$
    B. $(5y + 3)(2y - 3)$
    C. $(5y + 3)(2y - 1)$

18. $6n^8 - 11n^4 - 10$
    A. $(3n^4 - 5)(2n^4 + 2)$
    B. $(3n^4 + 2)(2n^4 - 5)$
    C. $(6n^4 - 5)(n^4 + 2)$

19. $nx^2 + x^2 - 9n - 9$
    A. $(x - 3)(x + 3)(n + 1)$
    B. $(x - 3)(x + 3)(n - 1)$
    C. $(x^2 + 1)(n - 9)$

20. $3x^{1/2} + 6x^{3/2}$

    A. $3x^{1/2}(1 + 2x^3)$

    B. $3x(x^{1/2} + 2x^3)$

    C. $3x^{1/2}(1 + 2x)$

In problems 21 through 23, solve for the value(s) of the variable.

21. $3x^2 - 28x + 32 = 0$

    A. 4 and 8

    B. -4 and $-\dfrac{8}{3}$

    C. 8 and $\dfrac{4}{3}$

22. $5x^2 - 3x - 1 = 0$

    A. 1 and $\dfrac{1}{5}$

    B. $\dfrac{3 \pm \sqrt{29}}{10}$

    C. $\dfrac{3 \pm \sqrt{11}}{10}$

23. $x^4 - 29x^2 + 100 = 0$

    A. 2, -2, 5, -5

    B. 10, -10

    C. 4, -4, 25, -25

In problems 24 and 25, determine all the values of the variable that satisfy the inequality.

24. $x^2 + 10x + 16 > 0$

    A. $x > 8$ or $x < -2$

    B. $x > 2$ or $x < -8$

    C. $x > -2$ or $x < -8$

25. $\dfrac{z-1}{z+4} \geq 0$

    A. $z < -4$ or $z \geq 1$

    B. $z \leq -4$ or $z > 1$

    C. $z < -1$ or $z \geq 4$

26. Solve for all values of x that satisfy the equation $\sqrt{5x + 9} + 1 = x$.

    A. 8 or -1

    B. -8 or 1

    C. 8 only

27. Determine the domain and range for the function $f(x) = \sqrt{x-2} + 1$.

    A. **Domain:** all real numbers; **range:** all positive numbers and 0
    B. **Domain:** all positive numbers; **range:** all numbers greater than 1
    C. **Domain:** all numbers greater than or equal to 2; **range:** all numbers greater than or equal to 1

28. If $f(x) = 4x^2 + 3x + 1$, then $f(-2) =$

    A. 59
    B. 11
    C. 10

29. Given the functions $f(x) = x^2 - x$ and $g(x) = 2x + x^2$, then $f + g =$

    A. $x^2 + x$
    B. $x^4 + x$
    C. $2x^2 + x$

30. Given the functions $f(x) = x^2 - x$ and $g(x) = 2x + x^2$, then $f - g =$

    A. $-3x$
    B. $x^2 - 3x$
    C. $2x^2 - 3x$

31. Given the functions $f(x) = x^2 - x$ and $g(x) = 2x + x^2$, then $f \cdot g =$

    A. $x^4 + x^3 - 2x^2$
    B. $2x^4 - x^3 + 2x^2$
    C. $x^4 - x^3 - 2x^2$

32. Given the functions $f(x) = x^2 - x$ and $g(x) = 2x + x^2$, then $\dfrac{f}{g} =$

    A. $-\dfrac{1}{2}$
    B. $\dfrac{x-1}{x+2}$
    C. $\dfrac{1-x}{2}$

33. Given the functions $f(x) = x^2 - x$ and $g(x) = 2x + x^2$, then the composition of functions, $f \circ g =$

    A. $x^4 + x^3 - 2x^2$
    B. $x^4 - 2x^3 + 3x^2 - 2x$
    C. $x^4 + 4x^3 + 3x^2 - 2x$

34. If $f(x) = 2x^2 - 3x + 4$, then what is $\dfrac{f(x+h) - f(x)}{h}$?

   A. 1

   B. $4x + 2h - 3$

   C. $4x - 2h + 3$

35. If $f(x) = 3x^3 - 4$, then what is the inverse function, $f^{-1}(x)$?

   A. $\dfrac{1}{3x^3 - 4}$

   B. $-3x^3 + 4$

   C. $\sqrt[3]{\dfrac{x+4}{3}}$

36. Given the function $y = x^3$, which of the following moves this function up 3 units and left 2 units?

   A. $y = (x + 2)^3 + 3$

   B. $y = (x - 2)^3 + 3$

   C. $y = (x + 3)^3 - 2$

37. Given the function $y = \sqrt{x}$, which of the following moves the function down 2 units and flips (reflects) it over a horizontal line?

   A. $y = -\sqrt{x} - 2$

   B. $y = -\sqrt{x - 2}$

   C. $y = \sqrt{-x} - 2$

38. Given the function $f(x) = x^6 + 3x^5 - 9x^4 - 4x^2 + x - 2$, use the Remainder Theorem to evaluate $f(2)$.

   A. 0

   B. 16

   C. -2

39. Find all the solutions of $x^3 - 2x^2 - 11x + 12 = 0$.

   A. $x = -1, x = 3, x = -4$

   B. $x = 1, x = -3, x = -4$

   C. $x = 1, x = -3, x = 4$

40. Find all the solutions of $x^4 + 2x^3 - 8x^2 - 18x - 9 = 0$.

   A. $x = -1, x = 1, x = 3, x = -3$

   B. $x = -1, x = 3, x = -3$

   C. $x = -1, x = 3$

41. Find the integer that is the lowest upper bound for the roots of $x^3 + 2x^2 + 3x - 4 = 0$.

 A. 1
 B. 3
 C. 4

42. Find the integer that is the highest lower bound for the roots of $x^4 - 3x^2 + 2x - 4 = 0$.

 A. −2
 B. −3
 C. −4

43. The graph of $2x - 3y = 8$ has an x-intercept of what?

 A. 4
 B. −4
 C. $-\dfrac{8}{3}$

44. The vertex of the parabola $y = x^2 - 4x + 7$ is in which quadrant?

 A. Quadrant I
 B. Quadrant III
 C. Quadrant IV

45. The graph of $y = (x - 1)(x + 3)(x - 4)$ has a y-intercept of what?

 A. −12
 B. 12
 C. 0

46. At which of the x values is the graph of $y = x^4 - 3x^3 + 2x^2 - 100$ above the x-axis?

 A. $x = -3$
 B. $x = 0$
 C. $x = 4$

47. The graph of which of these functions rises as it moves from left to right?

 A. $y = \left(\dfrac{3}{4}\right)^x$
 B. $y = 1.1^x$
 C. $y = 4^{-x}$

In problems 48 through 50, simplify the expression.

48. $\dfrac{e^{3x}}{e^{-x}}$

    A. e^{4x}

    B. e^{2x}

    C. e^{3}

49. $\left(e^{2x}\right)^{2} e^{3x}$

    A. $e^{4x^{2}+3x}$

    B. $e^{12x^{3}}$

    C. e^{7x}

50. $\left(e^{2x-1}\right)^{-2}$

    A. e^{-4x+2}

    B. $e^{1-4x+4x^{2}}$

    C. $\dfrac{1}{e^{(2x-1)^{2}}}$

51. At the end of five years, what is the total amount of an investment of $5,000 earning interest at the rate of 4.25 percent compounded quarterly?

    A. $6,062.60

    B. $6,156.73

    C. $6,176.90

52. At the end of 10 years, what is the total amount of an investment of $1,000 earning interest at the rate of 6 percent compounded continuously?

    A. $1,822.12

    B. $4,034.29

    C. $1,790.85

53. $\log_{2}8 =$

    A. 4

    B. 3

    C. 2

54. $\log_{4}\sqrt{2} =$

    A. $\dfrac{1}{4}$

    B. $\dfrac{1}{2}$

    C. 2

55. Simplify the expression $\log_3 \frac{x}{9}$.

   A. $-2 + \log_3 x$

   B. $2 \cdot \log_3 x$

   C. $\frac{\log_3 x}{2}$

56. How long will it take for a bacterial colony to double in size if it is growing at the rate of 2 percent per hour?

   A. About 2 hours

   B. About 35 hours

   C. About 67 hours

In problems 57 through 59, solve for the value of x.

57. $8^{2x+1} = 4^{x-1}$

   A. -2

   B. $-\frac{1}{2}$

   C. $-\frac{5}{4}$

58. $\log_2 x + \log_2(2x + 1) = \log_2 3$

   A. -1

   B. 1

   C. $\frac{3}{2}$

59. $\log(3x - 2) = 1$

   A. 1

   B. 4

   C. 10

60. Which is the y-intercept of the graph of $y = -x^4 + 3x^3 - 45x^2 + 13x + 1$?

   A. $(0,1)$

   B. $(0,-1)$

   C. $(1,-1)$

61. The graph of $y = (x - 14)^4 - 3$ has a minimum value of what?

   A. -3

   B. -14

   C. $-38,419$

62. The graph of $y = 4 - 3x^2 - x^3 - x^5$ does what as x gets very large?

    A. The y values get very small.

    B. The y values get very large.

    C. The y values get close to 0.

63. The graph of $y = \dfrac{x-3}{x+2}$ has a vertical asymptote of what?

    A. $x = 1$

    B. $x = 3$

    C. $x = -2$

64. The graph of $y = \dfrac{x^2 - 1}{2x^2 + 3x - 2}$ has a horizontal asymptote of what?

    A. $y = 1$

    B. $y = \dfrac{1}{2}$

    C. $y = 2$

65. The graph of $y = \sqrt[3]{x+1}$ crosses the y-axis where?

    A. $(-1, 0)$

    B. $(0, 1)$

    C. $(0, -1)$

66. The graph of $y = |2x + 3|$ has a minimum point when x is equal to what?

    A. 0

    B. $-\dfrac{3}{2}$

    C. $-\dfrac{2}{3}$

67. The graph of $y = \log_3 x$ crosses the x-axis at which point?

    A. $(1, 0)$

    B. $(0, 0)$

    C. $\left(\dfrac{1}{3}, 0\right)$

68. Rationalizing $\dfrac{8x}{\sqrt{10}}$ you get what?

    A. $\dfrac{\sqrt{10}}{8x}$

    B. $\dfrac{4x\sqrt{10}}{5}$

    C. $\dfrac{\sqrt{10}}{80x}$

69. Rewrite the following without a radical in the denominator: $\dfrac{4}{\sqrt{3}-x}$.

    A. $\dfrac{4\left(\sqrt{3}+x\right)}{3-x^2}$

    B. $\dfrac{4\left(\sqrt{3}-x\right)}{9-x^2}$

    C. $\dfrac{4\left(\sqrt{3}-x\right)}{3-x}$

70. Solve for x in the equation $\sqrt{5x+1}=x-1$.

    A. 0

    B. 7

    C. 60

71. Solve for x in the equation $\sqrt{2x+3}-\sqrt{x-10}=4$.

    A. 11

    B. 11 and 59

    C. 21

In problems 72 through 78, rewrite the expression in the complex form $a + bi$.

72. $\sqrt{-25}$

    A. $5i$

    B. $5 + 5i$

    C. $25i$

73. $\sqrt{6-4(14)}$

    A. $6 - 56i$

    B. $50i$

    C. $5\sqrt{2}\,i$

74. $\dfrac{3\pm\sqrt{9-4(5)}}{10}$

    A. $\dfrac{3}{10}\pm\sqrt{11}i$

    B. $\dfrac{3}{10}\pm\dfrac{\sqrt{11}}{10}i$

    C. $3i\pm\dfrac{\sqrt{11}}{10}$

75. $(8 - 3i) + (4 + 2i)$

   A. $12 - i$

   B. $4 - 5i$

   C. $4 - i$

76. $(8 - 3i) - (4 + 2i)$

   A. $12 - i$

   B. $4 - 5i$

   C. $4 - i$

77. $(8 - 3i)(4 + 2i)$

   A. $32 - 6i$

   B. $26 + 4i$

   C. $38 + 4i$

78. $\dfrac{8 - 3i}{4 + 2i}$

   A. $\dfrac{13}{10} - \dfrac{7}{5}i$

   B. $\dfrac{13}{6} - \dfrac{7}{3}i$

   C. $\dfrac{19}{10} - \dfrac{7}{5}i$

79. Solve for the value(s) of x in the equation $x^2 + 3x + 7 = 0$.

   A. -4 and 3

   B. $-\dfrac{3}{2} \pm \dfrac{\sqrt{19}}{2}i$

   C. $-\dfrac{3}{2} \pm \dfrac{5}{2}i$

80. Solve for the value(s) of x in the equation $x^2 + 81 = 0$.

   A. 9 and -9

   B. $9i$ and $-9i$

   C. $\dfrac{9}{2}i$ and $-\dfrac{9}{2}i$

81. Solve for the value(s) of x in the equation $\dfrac{3x}{x - 2} - \dfrac{2x}{x + 5} = \dfrac{3x^2 + 6x + 15}{(x - 2)(x + 5)}$.

   A. 5

   B. 5 and 1

   C. 5 and $\dfrac{3}{2}$

82. If y varies directly with the square of x, and y equals 8 when x equals 4, then what is y when x is 10?

    A. 5
    B. 20
    C. 50

83. If y varies inversely with x, and y equals 10 when x equals 2, then what is y when x equals 4?

    A. 5
    B. 20
    C. 100

84. If the radius of a circle is 7 inches, then what is its circumference?

    A. 7π
    B. 14π
    C. 49π

85. If the diameter of a circle is 30 cm., then what is its area in square centimeters?

    A. 900π
    B. 225π
    C. 60π

86. What is the center of the circle $(x - 3)^2 + (y + 4)^2 = 25$?

    A. $(3,-4)$
    B. $(-3,4)$
    C. $(3,4)$

87. What is the radius of the circle $(x - 3)^2 + (y + 4)^2 = 25$?

    A. 5
    B. 25
    C. 625

88. The standard form of the circle $x^2 + y^2 - 6x + 10y + 30 = 0$ is which of the following?

    A. $(x - 6)^2 + (y + 10)^2 = 30$
    B. $(x - 3)^2 + (y + 5)^2 = 4$
    C. $(x + 6)^2 + (y - 10)^2 = 30$

89. The ellipse $\dfrac{(x-3)^2}{9} + \dfrac{(y+5)^2}{25} = 1$ has which of the following properties?

   A. Its center is at $(3,-5)$, and it is wider than it is tall.

   B. Its center is at $(3,-5)$, and it is taller than it is wide.

   C. Its center is at $(-3,5)$, and it is wider than it is tall.

90. The standard form for the ellipse $16x^2 + y^2 + 32x - 6y + 9 = 0$ is which of the following?

   A. $\dfrac{(x+1)^2}{1} + \dfrac{(y-3)^2}{16} = 1$

   B. $\dfrac{(x+1)^2}{16} + \dfrac{(y-3)^2}{1} = 1$

   C. $\dfrac{(x-1)^2}{1} + \dfrac{(y+3)^2}{16} = 1$

91. A parabola with a vertex of $(-1,3)$ that opens downward is which of the following?

   A. $y = (x+1)^2 - 3$

   B. $y = -(x+1)^2 - 3$

   C. $y = -(x+1)^2 + 3$

92. The vertex of the parabola $x^2 - 6x + y + 5 = 0$ is which of the following?

   A. $(3,-4)$

   B. $(3,4)$

   C. $(-3,4)$

93. The hyperbola $\dfrac{(x-1)^2}{16} - \dfrac{y^2}{36} = 1$ has which of the following characteristics?

   A. Its center is at $(1,0)$, and it opens left and right.

   B. Its center is at $(-1,0)$, and it opens left and right.

   C. Its center is at $(1,0)$, and it opens upward and downward.

94. Which of the following is the equation of all the circles with their centers at $(-1,3)$?

   A. $(x-1)^2 + (y+3)^2 = r^2$

   B. $(x-1)^2 + (y+3)^2 = 1$

   C. $(x+1)^2 + (y-3)^2 = r^2$

95. Which of the following are the equations of all the circles with their centers on the y-axis and with a radius of 5?

   A. $(x-a)^2 + y^2 = 25$

   B. $x^2 + (y-b)^2 = 25$

   C. $x^2 + (y-5)^2 = r^2$

96. Solve the system of equations using the addition method: $\begin{array}{l} 5x - 2y = 7 \\ 3x + 2y = 9 \end{array}$

    A. $x = 2, \ y = \dfrac{3}{2}$

    B. $x = 5, y = 9$

    C. $x = -1, y = 6$

97. Solve the system of equations using the addition method: $\begin{array}{l} x + 3y = 1 \\ 3x - y = 13 \end{array}$

    A. $x = 7, y = -2$

    B. $x = 4, y = -1$

    C. $x = -1, y = -11$

98. Solve the system of equations using substitution: $\begin{array}{l} x - 5y = 13 \\ 3x - 4y = 6 \end{array}$

    A. $x = -12, y = -5$

    B. $x = 2, y = 0$

    C. $x = -2, y = -3$

99. Solve the system of equations using Cramer's Rule: $\begin{array}{l} 4x + 3y = 8 \\ 5x - 2y = 11 \end{array}$

    A. $x = \dfrac{49}{23} \ y = -\dfrac{84}{23}$

    B. $x = \dfrac{49}{23} \ y = -\dfrac{4}{23}$

    C. $x = -\dfrac{49}{23} \ y = \dfrac{84}{23}$

100. Solve the system of equations: $\begin{array}{l} x^2 + y^2 = 29 \\ 2x^2 - y^2 = 46 \end{array}$

    A. $(5,2), (-5,2), (5,-2), (-5,-2)$

    B. $\left(\sqrt{28}, 1\right), \left(\sqrt{28}, -1\right), \left(-\sqrt{28}, 1\right), \left(-\sqrt{28}, -1\right)$

    C. $\left(\sqrt{23}, 0\right), \left(-\sqrt{23}, 0\right)$

101. Solve the system of equations: $\begin{array}{l} 2x^2 - 3y^2 = 15 \\ x + 2y = 5 \end{array}$

    A. $(3,1), (-9,7)$

    B. $(5,1), (-9,7), (-5,1), (9,-7)$

    C. $(-5,1), (9,-7), (5,1), (9,7)$

102. The sum of the squares of two positive numbers is 74, and the difference of their squares is 24. What is the smaller of the two numbers?

    A. 4

    B. 5

    C. 6

103. Which of the points listed here belongs in the solution of the system of inequalities?

$x^2 + y^2 > 25$
$x + y < 2$

A. (0,0)

B. (−5,−3)

C. (−4,1)

104. Which of the points listed here is the solution of the system of equations?

$2x + 3y − 4z = 9$
$4x − 2y + z = −15$
$3x + 4y − z = 7$

A. $x = 1, y = 1, z = −1$

B. $x = 2, y = 10, z = −3$

C. $x = −2, y = 3, z = −1$

105. The system of equations shown here has a solution with an x value of what?

$x − y + 2z + 3w = 17$
$2x − y − z + 4w = 24$
$3x + 2y − 3z + w = 9$
$2x + 3y + z + 5w = 20$

A. 1

B. 3

C. 4

106. If matrix A = $\begin{bmatrix} −1 & 4 \\ 9 & 0 \\ −3 & 2 \end{bmatrix}$ and if matrix B = $\begin{bmatrix} 4 & −2 \\ 3 & 4 \\ 0 & −1 \end{bmatrix}$, then A + B =

A. $\begin{bmatrix} 3 & 2 \\ 12 & 4 \\ −3 & 1 \end{bmatrix}$

B. $\begin{bmatrix} −5 & 6 \\ 6 & −4 \\ −3 & −3 \end{bmatrix}$

C. $\begin{bmatrix} −3 & −2 \\ −12 & −4 \\ 3 & −1 \end{bmatrix}$

107. If matrix C = $\begin{bmatrix} 2 & −3 \\ 4 & −5 \end{bmatrix}$ and if matrix D = $\begin{bmatrix} −1 & −4 \\ 2 & 7 \end{bmatrix}$, then C·D =

A. $\begin{bmatrix} −8 & −29 \\ −14 & −51 \end{bmatrix}$

B. $\begin{bmatrix} −2 & 12 \\ 8 & −35 \end{bmatrix}$

C. $\begin{bmatrix} 2 & −12 \\ −8 & 35 \end{bmatrix}$

108. If matrix D $= \begin{bmatrix} -1 & -4 \\ 2 & 7 \end{bmatrix}$, then the inverse of D, $D^{-1} =$

   A. $\begin{bmatrix} 1 & 4 \\ -2 & -7 \end{bmatrix}$

   B. $\begin{bmatrix} 2 & 7 \\ -1 & -4 \end{bmatrix}$

   C. $\begin{bmatrix} 7 & 4 \\ -2 & -1 \end{bmatrix}$

109. Alex bought 4 apples, 2 bananas, and 3 cantaloupes for $4.55. Betty bought 10 apples and 4 cantaloupes for $8. Carla bought 1 apple, 6 bananas, and 1 cantaloupe for $2.15. How much did each of the apples cost?

   A. $.40

   B. $.50

   C. $.75

Key to Pretest Questions

   1. B

   2. A

   3. C

If you missed 1, 2, or 3, go to "Rules for Exponents," page 24.

   4. A

   5. A

If you missed 4 or 5, go to "Adding and Subtracting Polynomials," page 26.

   6. B

   7. C

If you missed 6 or 7, go to "Multiplying Polynomials," page 28.

   8. C

   9. B

   10. C

   11. A

If you missed 8, 9, 10, or 11, go to "Special Products," page 30.

12. A

13. B

If you missed 12 or 13, go to "Dividing Polynomials," page 35.

14. A

If you missed 14, go to "Greatest Common Factor," page 41.

15. C

16. B

If you missed 15 or 16, go to "Factoring Binomials," page 44.

17. A

18. B

If you missed 17 or 18, go to "Factoring Trinomials," page 46.

19. A

If you missed 19, go to "Factoring by Grouping," page 50.

20. C

If you missed 20, go to "Greatest Common Factor," page 41.

21. C

22. B

If you missed 21 or 22, go to "Solving Quadratic Equations," page 52.

23. A

If you missed 23, go to "Solving Quadratic-Like Equations," page 63.

24. C

25. A

If you missed 24 or 25, go to "Quadratic and Other Inequalities," page 66.

26. C

If you missed 26, go to "Radical Equations with Quadratics," page 70.

27. C

28. B

If you missed 27 or 28, go to "Functions and Function Notation," page 74.

29. C

30. A

31. A

32. B

If you missed 29, 30, 31, or 32, go to "Function Operations," page 77.

33. C

34. B

If you missed 33 or 34, go to "Composition of Functions," page 79, and "Difference Quotient," page 80.

35. C

If you missed 35, go to "Inverse Functions," page 82.

36. A

37. A

If you missed 36 or 37, go to "Function Transformations," page 85.

38. A

If you missed 38, go to "Remainder Theorem," page 90.

39. C

40. B

If you missed 39 or 40, go to "Rational Root Theorem," page 92.

41. A

42. B

If you missed 41 or 42, go to "Upper and Lower Bounds," page 96.

43. A

44. A

45. B

46. A

47. B

If you missed 43, 44, 45, 46, or 47, go to "Using a Graphing Calculator to Graph Lines and Polynomials," page 98.

48. A

49. C

50. A

If you missed 48, 49, or 50, go to "The Constant 'e'," page 112.

51. C

52. A

If you missed 51 or 52, go to "Compound Interest," page 114.

53. B

54. A

If you missed 53 or 54, go to "Logarithmic Functions," page 117.

55. A

If you missed 55, go to "Laws of Logarithms," page 119.

56. B

If you missed 56, go to "Applications of Logarithms," page 121.

57. C

58. B

59. B

If you missed 57, 58, or 59, go to "Solving Exponential and Logarithmic Equations," page 124.

60. A

61. A

62. A

If you missed 60, 61, or 62, go to "Graphing Polynomials," page 130.

63. C

64. B

If you missed 63 or 64, go to "Graphing Rational Functions," page 135.

65. B

If you missed 65, go to "Graphing Radical Functions," page 139.

66. B

67. A

If you missed 66 or 67, go to "Graphing Absolute Value and Logarithmic Functions," page 144.

68. B

69. A

If you missed 68 or 69, go to "Radical Equations and Conjugates," page 157.

70. B

71. B

If you missed 70 or 71, go to "Radical Equations—Squaring More than Once," page 158.

72. A

73. C

74. B

If you missed 72, 73, or 74, go to "Complex Numbers," page 160.

75. A

76. B

77. C

78. A

If you missed 75, 76, 77, or 78, go to "Operations Involving Complex Numbers," page 162.

79. B

80. B

If you missed 79 or 80, go to "Quadratic Formula and Complex Numbers," page 164.

81. C

If you missed 81, go to "Rational Equations," page 166.

82. C

83. A

If you missed 82 or 83, go to "Variation," page 168.

84. B

85. B

86. A

If you missed 84, 85, or 86, go to "Circle," page 171.

87. A

88. B

If you missed 87 or 88, go to "Circle/Standard Form," page 171.

89. B

90. A

If you missed 89 or 90, go to "Ellipse," page 174.

91. C

92. B

If you missed 91 or 92, go to "Parabola," page 177.

93. A

If you missed 93, go to "Hyperbola," page 179.

94. C

95. B

If you missed 94 or 95, go to "Writing Equations of Circles," page 190.

96. A

97. B

If you missed 96 or 97, go to "Solving Linear Systems Using the Addition Method," page 193.

98. C

If you missed 98, go to "Solving Linear Equations Using Substitution," page 195.

99. B

If you missed 99, go to "Solving Linear Equations Using Cramer's Rule," page 196.

100. A

101. A

If you missed 100 or 101, go to "Systems of Non-Linear Equations," page 199.

102. B

If you missed 102, go to "Story Problems Using Systems of Equations," page 201.

103. B

104. C

105. B

If you missed 103, 104, or 105, go to "Systems of Inequalities," page 204.

106. A

107. A

If you missed 106 or 107, go to "Operations on Matrices," page 220.

108. C

If you missed 108, go to "Inverse of a 2×2 Matrix," page 223.

109. B

If you missed 109, go to "Applications of Systems of Linear Equations," page 225.

Chapter 1
The Basics

Algebra is a language. You need to know the rules and definitions to understand this language and its many manipulations. In this chapter is a review of some of the important basics of algebra: rules for exponents and operations involving polynomials. These should be reviewed before going on to some of the advanced topics in Algebra II.

Rules for Exponents

A power or exponent tells how many times a number multiplies itself. Many opportunities exist in algebra for combining and simplifying expressions with two or more of these exponential terms in them. The rules used here to combine numbers and variables work for any expression with exponents. They are found in formulas and applications in science, business, and technology, as well as math. The term a^4 has an exponent of 4 and a *base* of a. The base is what gets multiplied repeatedly. The exponent tells how many times that *repeatedly* is.

Laws for Using Exponents

$a^n \cdot a^m = a^{n+m}$ When multiplying two numbers that have the same base, add their exponents.

$\dfrac{a^n}{a^m} = a^{n-m}$ When dividing two numbers that have the same base, subtract their exponents.

$(a^n)^m = a^{n \cdot m}$ When raising a value that has an exponent to another power, multiply the two exponents.

$(a \cdot b)^n = a^n \cdot b^n$ The product of two numbers raised to a power is equal to raising each number to that power and then multiplying them together.

$\left(\dfrac{a}{b}\right)^n = \dfrac{a^n}{b^n}$ The quotient of two numbers raised to a power is equal to raising each of the numbers to that power and then dividing them.

$a^{-n} = \dfrac{1}{a^n}$ A value raised to a negative power can be written as a fraction with the positive power of that number in the denominator.

$a^0 = 1$ Any number (except 0) raised to the 0 power is equal to 1.

Example Problems

These problems show the answers and solutions.

1. Simplify: $x^4\left(\dfrac{x^3}{x^{-2}}\right)^3$

   answer: x^{19}

   In this case, the course of action is to simplify the expressions inside the parentheses first, raise that result to the third power, and finally multiply by the first factor.

   $$x^4\left(\frac{x^3}{x^{-2}}\right)^3 = x^4\left(x^3 \cdot x^2\right)^3 = x^4\left(x^5\right)^3 = x^4\left(x^{15}\right) = x^{19}$$

2. Simplify: $y^{-3}\left(y^2\right)^4 \cdot \dfrac{y^4}{yy^3}$

   answer: y^5

   The denominator reads yy^3, which implies that the first factor has an exponent of 1, reading y^1y^3.

   $$y^{-3}\left(y^2\right)^4 \cdot \frac{y^4}{yy^3} = y^{-3} \cdot y^8 \cdot \frac{y^4}{y^4} = y^5 \cdot 1 = y^5$$

3. Simplify: $\left(\dfrac{a^4 b^2}{a^{-1} b^4}\right)^5 \cdot \left(\dfrac{a^3}{ab^4}\right)^{-1}$

   answer: $\dfrac{a^{23}}{b^6}$

   A nice property of fractions is that when they're raised to a negative power, you can rewrite the expression and change the power to a positive if you "flip" the fraction. So $\left(\dfrac{a^3}{ab^4}\right)^{-1} = \left(\dfrac{ab^4}{a^3}\right)^1$. First, rewrite the second fraction without the negative exponent. Then simplify the fractions inside the parentheses. The next step is to raise the factors in the parentheses to the powers. Lastly, multiply the terms in the two numerators and denominators.

   $$\left(\frac{a^4 b^2}{a^{-1} b^4}\right)^5\left(\frac{a^3}{ab^4}\right)^{-1} = \left(\frac{a^4 b^2}{a^{-1} b^4}\right)^5\left(\frac{ab^4}{a^3}\right)^1 = \left(\frac{a^4 a^1}{b^2}\right)^5\left(\frac{b^4}{a^2}\right)^1 = \left(\frac{a^5}{b^2}\right)^5\left(\frac{b^4}{a^2}\right)^1 = \frac{a^{25} \cdot b^4}{b^{10} \cdot a^2} = \frac{a^{25}}{b^{6}} \cdot \frac{b^4}{a^2} = \frac{a^{23}}{b^6}$$

Work Problems

Use these problems to give yourself additional practice.

1. Simplify: $\left(\dfrac{x^3}{x^{-3}}\right)^2$

2. Simplify: $\left(\dfrac{y^4}{3x^2}\right)^{-3}$

3. Simplify: $\dfrac{a^{-2} x^2}{a^4 x} \cdot \left(\dfrac{a^3}{x^3}\right)^4$

4. Simplify: $\dfrac{\left(abc^2\right)^4}{a^2 bc^{-1}}$

5. Simplify: $\left(\dfrac{x^2 y}{2w^4 z}\right)^4\left(\dfrac{w^4 z^6}{xy}\right)^4$

Worked Solutions

1. x^{12} First simplify inside the parentheses. Then raise the result to the second power.

$$\left(\frac{x^3}{x^{-3}}\right)^2 = \left(x^6\right)^2 = x^{12}$$

2. $\dfrac{27x^6}{y^{12}}$ First "flip" the fraction and change the power to positive.

$$\left(\frac{y^4}{3x^2}\right)^{-3} = \left(\frac{3x^2}{y^4}\right)^3 = \frac{27x^6}{y^{12}}$$

3. $\dfrac{a^6}{x^{11}}$ First raise the factors in the parentheses to the fourth power. Then simplify the first fraction before multiplying the two fractions together.

$$\frac{a^{-2}x^2}{a^4 x}\cdot\left(\frac{a^3}{x^3}\right)^4 = \frac{a^{-2}x^2}{a^4 x}\cdot\frac{a^{12}}{x^{12}} = \frac{\cancel{x}}{a^6}\cdot\frac{a^{\cancel{12}\,6}}{x^{\cancel{12}\,11}} = \frac{a^6}{x^{11}}$$

4. $a^2b^3c^9$ First raise the numerator to the fourth power. Then simplify the fraction.

$$\frac{\left(abc^2\right)^4}{a^2bc^{-1}} = \frac{a^4b^4c^8}{a^2bc^{-1}} = \frac{a^{\cancel{4}\,2}b^{\cancel{4}\,3}c^{\cancel{8}\,9}}{\cancel{a^2}\,\cancel{b}\,\cancel{c^{-1}}} = a^2b^3c^9$$

5. $\dfrac{x^4 z^{20}}{16}$ Since both fractions are raised to the fourth power, it is easier to combine them in the same parentheses and then later raise the result to the fourth power.

$$\left(\frac{x^2 y}{2w^4 z}\right)^4\left(\frac{w^4 z^6}{xy}\right)^4 = \left(\frac{x^2 y}{2w^4 z}\cdot\frac{w^4 z^6}{xy}\right)^4 = \left(\frac{x^{\cancel{2}\,1}\,\cancel{y}}{2w^4\,\cancel{z}}\cdot\frac{\cancel{w^4}\,z^{\cancel{6}\,5}}{\cancel{x}\,\cancel{y}}\right)^4 = \left(\frac{xz^5}{2}\right)^4 = \frac{x^4 z^{20}}{16}$$

Adding and Subtracting Polynomials

One major objective of working with algebraic expressions is to write them as simply as possible and in a logical, generally accepted arrangement. When more than one term exists (a term consists of one or more factors multiplied together and separated from other terms by + or −), then you check to see whether they can be combined with other terms that are *like* them. Numbers by themselves without letters or variables are *like* terms. You can combine 14 and 8 because you know what they are and know the rules. For instance, $14 + 8 = 22$, $14 - 8 = 6$, $14(8) = 112$, and so on. Numbers can be written so they can combine with one another. They can be added, subtracted, multiplied, and divided, as long as you don't divide by zero. Fractions can be added if they have a common denominator. Algebraic expressions involving variables or letters have to be dealt with carefully. Since the numbers that the letters represent aren't usually known, you can't add or subtract terms with different letters. The expression $2a + 3b$ has to stay that way. That's as simple as you can write it, but the expression $4c + 3c$ can be simplified. You don't know what c represents, but you can combine the terms to tell how many of them you have (even though you don't know what they are!): $4c + 3c = 7c$. Here are some other examples:

$$5ab + 9ab = 14ab$$
$$5x^2y - x^2y + 6xy^2 + 2xy^2 = 4x^2y + 8xy^2$$

Notice that there are two different kinds of terms, one with the x squared and the other with the y squared. Only those that have the letters exactly alike with the exact same powers can be combined. The only thing affected by adding and subtracting these terms is the coefficient.

Example Problems

These problems show the answers and solutions.

1. Simplify $5xy^2 + 8x - 9y^2 - 2x + 3y^2 - 8xy^2$.

   answer: $-3xy^2 + 6x - 6y^2$

   There are three different kinds of terms. First rearrange the terms so that the like terms are together. Then combine the like terms.

   $$5xy^2 + 8x - 9y^2 - 2x + 3y^2 - 8xy^2 = 5xy^2 - 8xy^2 + 8x - 2x - 9y^2 + 3y^2 = -3xy^2 + 6x - 6y^2$$

2. Simplify $3x^4 - 2x^3 + x^2 - x + 5 - 2x^2 + 3x^3 + 11$.

   answer: $3x^4 + x^3 - x^2 - x + 16$

   Rearrange the terms so that the like ones are together. By convention, you write terms that have different powers of the same variable in either decreasing or increasing order of their powers.

   $$3x^4 - 2x^3 + x^2 - x + 5 - 2x^2 + 3x^3 + 11 = 3x^4 - 2x^3 + 3x^3 + x^2 - 2x^2 - x + 5 + 11$$
   $$= 3x^4 + x^3 - x^2 - x + 16$$

Work Problems

Use these problems to give yourself additional practice.

1. Simplify by combining like terms: $m^2 + 3mn + m + 8 + 9mn - m^2 + 14m$.

2. Simplify by combining like terms: $3a + 4b - 6 + 2a - 11$.

3. Simplify by combining like terms: $2\pi r^2 + 8\pi r^2 - 6\pi r + 7$.

4. Simplify by combining like terms: $a + ab + ac + ad + ae + 1$.

5. Simplify by combining like terms: $5x^2y - 2x^2y + x^2y + xy^2 - 8y^2$.

Worked Solutions

1. **$12mn + 15m + 8$**   Rearrange the terms so that the terms that can be combined are together.

   $$m^2 + 3mn + m + 8 + 9mn - m^2 + 14m = m^2 - m^2 + 3mn + 9mn + m + 14m + 8$$
   $$= 12mn + 15m + 8$$

2. **$5a + 4b - 17$** $3a + 4b - 6 + 2a - 11 = 3a + 2a + 4b - 6 - 11 = 5a + 4b - 17$

3. **$10\pi r^2 - 6\pi r + 7$**   Only the terms with the πr^2 will combine.

4. **$a + ab + ac + ad + ae + 1$**   This is already simplified. None of the terms have exactly the same variables. There's nothing more to do.

5. **$4x^2y + xy^2 - 8y^2$**   There are three terms with the same variables raised to the same powers. Be very careful with problems like this.

Multiplying Polynomials

Multiplying polynomials requires that each term in one polynomial multiplies each term in the other polynomial. When a monomial (one term) multiplies another polynomial, the distributive property is used, and the result quickly follows. Multiplying polynomials with more than one term can be very complicated or tedious, but some procedures or methods can be used to provide better organization and accuracy. For instance, to multiply a binomial times another binomial, such as $(x + 2)(a + 3)$, or to multiply a binomial times a trinomial, such as $(x + 2)(x + y + 3)$, you can use the distributive property. Distribute the $(x + 2)$ over the other terms.

$$(x + 2)(a + 3) = (x + 2)(a) + (x + 2)(3)$$
$$= x(a) + 2(a) + x(3) + 2(3)$$
$$= ax + 2a + 3x + 6$$

$$(x + 2)(x + y + 3) = (x + 2)(x) + (x + 2)(y) + (x + 2)(3)$$
$$= x(x) + 2(x) + x(y) + 2(y) + x(3) + 2(3)$$
$$= x^2 + 2x + xy + 2y + 3x + 6$$
$$= x^2 + xy + 5x + 2y + 6$$

None of the terms are alike, so this can't be simplified further.

The FOIL Method

Multiplying two binomials together is a very common operation in algebra. The FOIL method is preferred when multiplying most types of binomials.

The letters in FOIL stand for First, Outer, Inner, and Last. These words describe the positions of the terms in the two binomials. Each term actually will have two different names, because each term is used twice in the process. In the multiplication problem, $(x + 2)(a + 3)$:

> The x and the a are the First terms of each binomial.
>
> The x and the 3 are the Outer terms in the two binomials.
>
> The 2 and the a are the Inner terms in the two binomials.
>
> The 2 and the 3 are the Last terms of each binomial.

These pairings tell you what to multiply.

> F: multiply $x \cdot a$
>
> O: multiply $x \cdot 3$
>
> I: multiply $2 \cdot a$
>
> L: multiply $2 \cdot 3$

Add these together, $x \cdot a + x \cdot 3 + 2 \cdot a + 2 \cdot 3 = ax + 3x + 2a + 6$. It's the same result, in a slightly different order, as the one obtained previously with distribution.

When using this FOIL method, you'll notice that, when the two binomials are alike—that is they have the same types of terms—the Outer and Inner terms combine, and the result is a trinomial. If they aren't alike, as shown by this last example, then none of the terms in the solution will combine, and you'll have a four-term polynomial.

Example Problems

These problems show the answers and solutions.

1. $(x + 3)(x - 8)$

    answer: $x^2 - 5x - 24$

    Using FOIL,

    F: multiply $x \cdot x$
    O: multiply $x(-8)$
    I: multiply $3 \cdot x$
    L: multiply $3(-8)$

    Add the products together: $x^2 - 8x + 3x - 24 = x^2 - 5x - 24$

2. $(3x - 1)(x - y)$

    answer: $3x^2 - 3xy - x + y$

    Using FOIL,

    F: multiply $3x \cdot x$
    O: multiply $3x(-y)$
    I: multiply $-1 \cdot x$
    L: multiply $-1(-y)$

    Add the products together: $3x^2 - 3xy - x + y$. Notice that none of the terms combine—they're not quite alike.

3. $(3x^2 - 1)(x^2 - 2)$

    answer: $3x^4 - 7x^2 + 2$

    Using FOIL,

    F: multiply $3x^2 \cdot x^2$
    O: multiply $3x^2(-2)$
    I: multiply $-1 \cdot x^2$
    L: multiply $-1(-2)$

    Add the products together: $3x^4 - 6x^2 - x^2 + 2 = 3x^4 - 7x^2 + 2$

Work Problems

Use these problems to give yourself additional practice. Find the products and simplify the answers.

1. $(x - 3)(x + 7)$

2. $(2y + 3)(3y - 2)$

3. $(6x - 3)(6x - 5)$

4. $(ax - 1)(bx - 2)$

5. $(m^3 - 3)(m^3 + 11)$

Worked Solutions

1. $x^2 + 4x - 21$ Using FOIL, $(x - 3)(x + 7) = (x - 3)(x + 7) = x^2 + 7x - 3x - 21$. The middle two terms combine.

2. $6y^2 + 5y - 6$ Using FOIL, $(2y + 3)(3y - 2) = 6y^2 - 4y + 9y - 6$. Again, the middle two terms combine.

3. $36x^2 - 48x + 15$ Using FOIL, $(6x - 3)(6x - 5) = 36x^2 - 30x - 18x + 15$.

4. $abx^2 - (2a + b)x + 2$ At first, it appears that the middle two terms can't combine. If the a and b had been numbers, you could have added them together. They still can be added and the result multiplies the x: $(ax - 1)(bx - 2) = ax \cdot bx - 2ax - bx + 2 = abx^2 - (2a + b)x + 2$.

5. $m^6 + 8m^3 - 33$ Using FOIL, $(m^3 - 3)(m^3 + 11) = m^6 + 11m^3 - 3m^3 - 33$.

Special Products

It's nice to have the FOIL method to multiply two binomials together. Unfortunately, no other really handy tricks exist for multiplying other polynomials. Basically, you just distribute the smaller of the two polynomials over the other polynomial. Any polynomials can be multiplied together. The different types of multiplications are classified by the number of terms in the multiplier. Some products, however, are easier to perform because of patterns that exist in them. These patterns largely are due to the special types of polynomials that are being multiplied together. Whenever you can recognize a special situation and can take advantage of a pattern, you'll save time and be less likely to make an error. Here are the special products:

1. $(a + b)(a - b) = a^2 - b^2$ Multiplying the sum and difference of the same two numbers

2. $(a + b)^2 = a^2 + 2ab + b^2$ or   Squaring a binomial
   $(a - b)^2 = a^2 - 2ab + b^2$

3. $(a + b)^3 = a^3 + 3a^2b + 3ab^2 + b^3$ or   Cubing a binomial sum or difference
   $(a - b)^3 = a^3 - 3a^2b + 3ab^2 - b^3$

4. $(a + b)(a^2 - ab + b^2) = a^3 + b^3$ or
   $(a - b)(a^2 + ab + b^2) = a^3 - b^3$ Resulting in the sum or difference of two cubes

Special Product #1

$$(a + b)(a - b) = a^2 - b^2$$

This equation represents the product of two binomials that have the same two terms, but the terms are added in one binomial and subtracted in the other. It's better known as, "The sum and the difference of the same two numbers." Notice that if FOIL is used here, the Outer product, $-ab$, and the Inner product, ab, are the opposites of one another. This means that they add up to 0, leaving just the first term squared minus the last term squared.

Example Problems

These problems show the answers and solutions.

1. $(x + 8)(x - 8)$

    answer: $x^2 - 64$

    Just square the first and last terms and take the difference.

2. $(3x + 2z)(3x - 2z)$

    answer: $9x^2 - 4z^2$

    Again, we have a difference of two squares.

3. $(xyz + 2p^3)(xyz - 2p^3)$

    answer: $x^2y^2z^2 - 4p^6$

    $(xyz)^2 = x^2y^2z^2$ and $\left(2p^3\right)^2 = 4p^6$

Special Product #2

$$(a + b)^2 = a^2 + 2ab + b^2$$

This product is of a binomial times itself. It's known as, "The perfect square trinomial." The pattern is that the first and last terms in the trinomial are the squares of the two terms in the binomial. The middle term of the trinomial is twice the product of the two original terms. Because the first and last terms are squares, they'll always be positive. The sign of the middle term will depend upon whatever the operation is in the binomial.

Using FOIL on the square of a binomial, $(a + b)^2 = (a + b)(a + b) = a \cdot a + a \cdot b + b \cdot a + b \cdot b = a^2 + ab + ab + b^2 = a^2 + 2ab + b^2$.

Using the special product on $(a + b)^2$, first write down the squares of the two terms:

$$a^2 \quad b^2$$

Then double the product of the two terms, $2ab$. That's the middle term.

$$a^2 + 2ab + b^2$$

Example Problems

These problems show the answers and solutions.

1. $(y + 6)^2$

   answer: $y^2 + 12y + 36$

   The first term is the square of y. The middle term is twice the product of the two terms y and 6. The last term is the square of 6.

2. $(9z - 4)^2$

   answer: $81z^2 - 72z + 16$

   Notice the middle term has a minus sign.

3. $\left(m + \dfrac{1}{2}\right)^2$

   answer: $m^2 + m + \dfrac{1}{4}$

   The middle term is twice the product of m and $\dfrac{1}{2}$: $2 \cdot m \cdot \dfrac{1}{2} = m$.

Special Product #3

$$(a + b)^3 = a^3 + 3a^2b + 3ab^2 + b^3$$

or

$$(a - b)^3 = a^3 - 3a^2b + 3ab^2 - b^3$$

Methods exist for finding all powers of a binomial, such as $(a + b)^4$, $(a + b)^7$, $(a - b)^{30}$, and so on. Cubing a binomial is a common task, and the pattern in this special product helps. When the *sum* of two terms (a binomial) is cubed, there's a 1-3-3-1 pattern coupled with decreasing powers of the first term and increasing powers of the second term, and all the terms in the result are added. If there's a *difference* that's being cubed, the only change in the basic pattern is that the terms have alternating signs.

Example Problems

These problems show the answers and solutions.

1. $(1 + z)^3$

   answer: $(1 + z)^3 = 1 + 3z + 3z^2 + z^3$

   Using the 1-3-3-1 pattern as the base, the powers of 1 ($1^3, 1^2, 1^1, 1^0$) are placed with their decreasing powers, right after the numbers. Note that the 1^0 isn't written that way, but just as a 1. Then the increasing powers of z (z^0, z^1, z^2, z^3) are placed with the numbers and 1s. The terms are then simplified.

   $$1 \cdot 1^3 + 3 \cdot 1^2 \cdot z^1 + 3 \cdot 1^1 \cdot z^2 + 1 \cdot z^3 = 1 + 3z + 3z^2 + z^3$$

2. $(y - 3)^3$

answer: $y^3 - 9y^2 + 27y - 27$

This time the two terms have their zero-exponents for emphasis—especially on the number -3. This shows how the powers increase with each step. Now, simplify the expression.

$$
\begin{aligned}
\left(y - 3\right)^3 &= 1 \cdot y^3(-3)^0 + 3 \cdot y^2(-3)^1 + 3 \cdot y^1(-3)^2 + 1 \cdot y^0(-3)^3 \\
&= 1 \cdot y^3 \cdot 1 + 3 \cdot y^2(-3) + 3 \cdot y^1(9) + 1 \cdot 1(-27) \\
&= y^3 - 9y + 27y - 27
\end{aligned}
$$

As you see, the 1-3-3-1 pattern has disappeared, but using it to build the product was simpler than multiplying this out the "long way" with distribution.

Special Product #4

$$(a + b)(a^2 - ab + b^2) = a^3 + b^3$$

or

$$(a - b)(a^2 + ab + b^2) = a^3 - b^3$$

These products are paired because of their results. These are very specific types of products. The terms in the binomial and trinomial have to be just so. You wouldn't necessarily expect to see such strange combinations of factors to be multiplied except that these wonderful results do occur. Also, these combinations of multipliers will show up again when factoring binomials.

The first multiplier is a binomial. The second multiplier is a trinomial with the squares of the two terms in the binomial in the first and third positions. The middle term of the trinomial multiplier is the opposite of the product of the two terms in the binomial. The result is always the sum or difference of two cubes, and the operation between the two terms in the answer is the same as the operation in the original binomial.

Example Problems

These problems show the answers and solutions.

1. $(x + 2)(x^2 - 2x + 4)$

answer: $x^3 + 8$

Recognizing the pattern—the sum of the cube root of x^3 and the cube root of 8 multiplying the trinomial containing the squares of each of these roots and the opposite of their product—saves time. Here's how it looks if you have to multiply it all out:

$$(x + 2)(x^2 - 2x + 4) = (x + 2)(x^2) + (x + 2)(-2x) + (x + 2)(4)$$
$$= x \cdot x^2 + 2 \cdot x^2 + x(-2x) + 2(-2x) + x \cdot 4 + 2 \cdot 4 = x^3 + 2x^2 - 2x^2 - 4x + 4x + 8 = x^3 + 8$$

Notice how the four middle terms are opposites of one another and disappear. That's what always happens with this particular product.

2. $(3y - 1)(9y^2 + 3y + 1)$

answer: $27y^3 - 1$

The first term in the trinomial is the square of the first term in the binomial. The last term in the trinomial is the square of the second term in the binomial. The middle term is the opposite of the product of the two terms in the binomial. Since this fits the pattern, the product is the difference of the cubes of the two terms in the binomial.

Work Problems

Use these problems to give yourself additional practice. Find the products and simplify the answers.

1. $(p + 70)(p - 70)$

2. $(2z - 9)^2$

3. $(m + 4)^3$

4. $(5x - 1)^3$

5. $(8 - y)(64 + 8y + y^2)$

Worked Solutions

1.   **$p^2 - 4,900$**   This is the product of the sum and difference of the same two values. The result is always the difference of their squares.

2.   **$4z^2 - 36z + 81$**   This is a perfect square trinomial. It's the result of multiplying a binomial by itself. The first and last terms are the squares of the two terms in the binomial. The middle term is twice their product.

3.   **$m^3 + 12m^2 + 48m + 64$**   This is the cube of a binomial. The 1-3-3-1 pattern was used with decreasing powers of the first term and increasing powers of the second term.

$$(m + 4)^3 = 1 \cdot m^3 \cdot 4^0 + 3 \cdot m^2 \cdot 4^1 + 3 \cdot m^1 \cdot 4^2 + 1 \cdot m^0 \cdot 4^3 = m^3 + 12m^2 + 48m + 64$$

4.   **$125x^3 - 75x^2 + 15x - 1$**   This is also the cube of a binomial. This time the binomial is a difference, so there are alternating signs in the answer.

$$(5x - 1)^3 = 1 \cdot (5x)^3 \cdot (-1)^0 + 3 \cdot (5x)^2 \cdot (-1)^1 + 3 \cdot (5x)^1 \cdot (-1)^2 + 1 \cdot (5x)^0 \cdot (-1)^3$$
$$= 125x^3 - 75x^2 + 15x - 1$$

5.   **$512 - y^3$**   This is the special product that results in the difference between two perfect cubes. The first term in the binomial is 8, and its square, 64, is the first term in the trinomial. The second term in the binomial is $-y$, and its square is y^2. The middle term in the trinomial is the opposite of the product of the two terms in the binomial. So, rather than distributing the binomial over the trinomial, just write this special product as $8^3 - y^3 = 512 - y^3$.

Dividing Polynomials

The method used in the division of polynomials depends on the divisor—the term or terms that divide into the other expression. If the divisor has only one term, then the division can be done by splitting up the dividend (the expression being divided into) into separate terms and making a fraction of each term in the dividend with the divisor in the denominator. If the divisor has more than one term, however, the previous method does not work. You can use a shortcut with some special types of long division, but long division will always work.

Divisors with One Term

Where just one term is in the divisor, split the problem up into fractions and simplify each term.

Example Problems

These problems show the answers and solutions.

1. Divide: $(5y^5 - 10y^4 + 20y^2 - y + 15) \div (5y^2)$.

    answer: $y^3 - 2y^2 + 4 - \dfrac{1}{5y} + \dfrac{3}{y^2}$

    The divisor, $5y^2$, will be the denominator of each fraction.

    $$\frac{5y^5 - 10y^4 + 20y^2 - y + 15}{5y^2} = \frac{5y^5}{5y^2} - \frac{10y^4}{5y^2} + \frac{20y^2}{5y^2} - \frac{y}{5y^2} + \frac{15}{5y^2} = y^3 - 2y^2 + 4 - \frac{1}{5y} + \frac{3}{y^2}$$

2. Divide: $(6xy - 2x^2y^2 + 8xy^3) \div (2xy^2)$.

    answer: $\dfrac{3}{y} - x + 4y$

    The divisor will be the denominator of each fraction.

    $$\frac{6xy - 2x^2y^2 + 8xy^3}{2xy^2} = \frac{6xy}{2xy^2} - \frac{2x^2y^2}{2xy^2} + \frac{8xy^3}{2xy^2} = \frac{3}{y} - x + 4y$$

Long Division

When the divisor has two or more terms, you can do the operation with long division. This looks very much like the division done on whole numbers. It uses the same setup and operations.

Example Problems

These problems show the answers and solutions.

1. Divide using long division: $(4y^3 - 3y^2 + 2y - 7) \div (y - 2)$.

    answer: $\left(4y^3 - 3y^2 + 2y - 7\right) \div \left(y - 2\right) = 4y^2 + 5y + 12 + \dfrac{17}{y - 2}$

Write the dividend in decreasing powers, leaving spaces for any missing terms (skipped powers).

$$y - 2 \overline{\smash{\big)}\,4y^3 - 3y^2 + 2y - 7}$$

Focus on the first term in the divisor, the y. Then determine what must multiply that y so you get the first term in the dividend, the $4y^3$. Multiplying y by $4y^2$ will give you $4y^3$. Write the $4y^2$ above the $4y^3$.

$$\begin{array}{r} 4y^2 \\ y - 2 \overline{\smash{\big)}\,4y^3 - 3y^2 + 2y - 7} \end{array}$$

Now multiply both terms in the divisor by $4y^2$ and put the results under the terms in the dividend that are alike.

$$\begin{array}{r} 4y^2 \\ y - 2 \overline{\smash{\big)}\,4y^3 - 3y^2 + 2y - 7} \\ 4y^3 - 8y^2 \end{array}$$

Next, subtract. The easiest way is to change each term in the expression that you're subtracting to its opposite and add. The subtraction of signed numbers is done by changing the sign of the number being subtracted.

$$\begin{array}{r} 4y^2 \\ y - 2 \overline{\smash{\big)}\,4y^3 - 3y^2 + 2y - 7} \\ -4y^3 + 8y^2 \\ \hline + 5y^2 + 2y - 7 \end{array}$$

The remaining terms in the dividend are brought down and then the process is repeated. Determine what you have to multiply y by to get the new first term. Multiplying by $5y$ will do it.

$$\begin{array}{r} 4y^2 + 5y \\ y - 2 \overline{\smash{\big)}\,4y^3 - 3y^2 + 2y - 7} \\ -4y^3 + 8y^2 \\ \hline + 5y^2 + 2y - 7 \\ 5y^2 - 10y \end{array}$$

Again, change the signs and add.

$$\begin{array}{r} 4y^2 + 5y \\ y - 2 \overline{\smash{\big)}\,4y^3 - 3y^2 + 2y - 7} \\ -4y^3 + 8y^2 \\ \hline + 5y^2 + 2y - 7 \\ -5y^2 + 10y \\ \hline 12y - 7 \end{array}$$

You get the last part of the answer by multiplying the y in the divisor by 12 to get the new first term, the $12y$.

$$
\begin{array}{r}
4y^2 + 5y + 12 \\
y - 2 \overline{\smash{\big)}\, 4y^3 - 3y^2 + 2y - 7} \\
\underline{-4y^3 + 8y^2} \\
+ 5y^2 + 2y - 7 \\
\underline{-5y^2 + 10y} \\
12y - 7 \\
12y - 24
\end{array}
$$

Do the last subtraction. Whatever is left over is the remainder. The remainder is usually written as a fraction with the divisor in the denominator.

$$
\begin{array}{r}
4y^2 + 5y + 12 \\
y - 2 \overline{\smash{\big)}\, 4y^3 - 3y^2 + 2y - 7} \\
\underline{-4y^3 + 8y^2} \\
+ 5y^2 + 2y - 7 \\
\underline{-5y^2 + 10y} \\
12y - 7 \\
\underline{- 12y + 24} \\
17
\end{array}
$$

2. Use long division to divide $(8m^4 - 3m^2 + m - 12) \div (2m - 1)$.

answer: $4m^3 + 2m^2 - \dfrac{1}{2}m + \dfrac{1}{4} - \dfrac{11\frac{3}{4}}{2m - 1}$

In this problem, terms are missing. The m^3 term is missing in the decreasing powers.

$$
\begin{array}{r}
4m^3 + 2m^2 - \frac{1}{2}m + \frac{1}{4} \\
2m - 1 \overline{\smash{\big)}\, 8m^4 \qquad\;\; - 3m^2 + m - 12} \\
\underline{8m^4 - 4m^3} \\
4m^3 - 3m^2 + m - 12 \\
\underline{4m^3 - 2m^2} \\
- m^2 + m - 12 \\
\underline{- m^2 + \frac{1}{2}m} \\
\frac{1}{2}m - 12 \\
\underline{\frac{1}{2}m - \frac{1}{4}} \\
- 11\frac{3}{4}
\end{array}
$$

$$
(8m^4 - 3m^2 + m - 12) \div (2m - 1) = 4m^3 + 2m^2 - \frac{1}{2}m + \frac{1}{4} - \frac{11\frac{3}{4}}{2m - 1}
$$

Synthetic Division

Many division problems involving polynomials have special divisors of the form $x + a$ or $x - a$ where the coefficient of the variable is a 1. When this is the case, you can avoid long division and obtain the answer with a process called *synthetic division*. To do the problem $(3x^4 - 2x^3 + 91x + 11) \div (x + 3)$ using synthetic division, use the following steps:

1. Write just the coefficients of the terms in the dividend in decreasing order, inserting 0s for any missing terms (powers).

$$3 \quad -2 \quad 0 \quad 91 \quad 11$$

2. Put the opposite of the constant in the divisor in front of the row of coefficients.

$$-3 | 3 \ -2 \ 0 \ 91 \ 11$$

3. Draw a line two spaces below the row of coefficients and drop the first coefficient to below the line.

$$-3 | 3 \ -2 \ 0 \ 91 \ 11$$
$$\overline{ 3}$$

4. Multiply the constant in front times the dropped number and place the result below the second coefficient in the row. Add the two numbers together and put the result below the line.

$$-3 | 3 \ -2 \ 0 \ 91 \ 11$$
$$ -9$$
$$\overline{ 3 \ -11}$$

5. Multiply that result times the constant in front, put the result below the third coefficient in the row. Add the two numbers and put the result below the line. Repeat this process until you run out of numbers in the row of coefficients.

$$-3 | 3 \ -2 \ \ 0 \ \ 91 \ \ 11$$
$$ -9 \ \ 33 \ -99 \ \ 24$$
$$\overline{ 3 \ -11 \ 33 \ \ -8 \ \ 35}$$

6. Write the answer by using the new coefficients below the line and inserting powers of the variable, starting with a variable that has a power one less than the power of the dividend.

$$(3x^4 - 2x^3 + 91x + 11) \div (x + 3) = 3x^3 - 11x^2 + 33x - 8 + \frac{35}{x + 3}$$

In general, synthetic division is much preferred over long division. It takes less time and less space, and you make fewer errors in signs and computations. The operations used are multiplication and addition of the signed numbers. The constant in the divisor is changed to its opposite right at the beginning, taking care of the subtraction process. This next example shows more on inserting the 0s for missed terms.

Example Problems

1. Divide $(5x^6 - 3x^5 + 2x - 7) \div (x + 2)$.

    answer: $5x^5 - 13x^4 + 26x^3 - 52x^2 + 104x - 206 + \dfrac{405}{x+2}$

    Set up the problem by writing just the coefficients of the terms in a row. Put in zeros for any terms that are missing in the list of decreasing powers of the variable.

    $$
    \begin{array}{r|rrrrrrr}
    -2 & 5 & -3 & 0 & 0 & 0 & 2 & -7 \\
       &   & -10 & 26 & -52 & 104 & -208 & 412 \\
    \hline
       & 5 & -13 & 26 & -52 & 104 & -206 & 405
    \end{array}
    $$

    $$\left(5x^6 - 3x^5 + 2x - 7\right) \div (x + 2) = 5x^5 - 13x^4 + 26x^3 - 52x^2 + 104x - 206 + \frac{405}{x+2}$$

2. Divide $(4y^3 - 3y^2 + 2y - 7) \div (y - 2)$.

    answer: $4y^2 + 5y + 12 + \dfrac{17}{y-2}$

    Look at the example for long division on page 35. This is the same problem, except that it is done with synthetic division.

    $$
    \begin{array}{r|rrrr}
    +2 & 4 & -3 & 2 & -7 \\
       &   & 8 & 10 & 24 \\
    \hline
       & 4 & 5 & 12 & 17
    \end{array}
    $$

    Which do you prefer?

Work Problems

Use these problems to give yourself additional practice.

1. Use long division to divide $(8x^4 - 2x^2 + 18x + 6) \div (2x - 1)$.

2. Use long division to divide $(3x^3 + 22x^2 - 7x - 6) \div (3x - 2)$.

3. Use synthetic division to divide $(4x^4 - 2x^3 + 3x^2 - x - 8) \div (x + 2)$.

4. Use synthetic division to divide $(x^4 + 3x + 1) \div (x - 1)$.

5. Use synthetic division to divide $(x^9 + 27x^6 + x + 3) \div (x + 3)$.

Worked Solutions

1. $4x^3 + 2x^2 + 9 + \dfrac{15}{2x-1}$ Set up the long division problem. Focus on the first term of the divisor, the $2x$.

$$
\begin{array}{r}
4x^3 + 2x^2 \qquad\quad + 9 + \dfrac{15}{2x-1} \\
2x-1\,\overline{)\,8x^4 \qquad\quad - 2x^2 + 18x + 6} \\
\underline{8x^4 - 4x^3}\qquad\qquad\qquad \\
+\,4x^3 - 2x^2 + 18x + 6 \\
\underline{4x^3 - 2x^2}\qquad\qquad \\
18x + 6 \\
\underline{18x - 9} \\
15
\end{array}
$$

2. $x^2 + 8x + 3$

$$
\begin{array}{r}
x^2 + \; 8x + 3 \\
3x-2\,\overline{)\,3x^3 + 22x^2 - \; 7x - 6} \\
\underline{3x^3 - \; 2x^2}\qquad\qquad \\
24x^2 - \; 7x - 6 \\
\underline{24x^2 - 16x}\qquad \\
9x - 6 \\
\underline{9x - 6} \\
0
\end{array}
$$

In this case, there was no remainder. The binomial $3x - 2$ divided the expression evenly.

3. $4x^3 - 10x^2 + 23x - 47 + \dfrac{86}{x+2}$

$$
\begin{array}{r|rrrrr}
-2 & 4 & -2 & 3 & -1 & -8 \\
   &   & -8 & 20 & -46 & 94 \\
\hline
   & 4 & -10 & 23 & -47 & 86
\end{array}
$$

$$(4x^4 - 2x^3 + 3x^2 - x - 8) \div (x+2) = 4x^3 - 10x^2 + 23x - 47 + \dfrac{86}{x+2}$$

4. $x^3 + x^2 + x + 4 + \dfrac{5}{x-1}$

$$
\begin{array}{r|rrrrr}
1 & 1 & 0 & 0 & 3 & 1 \\
  &   & 1 & 1 & 1 & 4 \\
\hline
  & 1 & 1 & 1 & 4 & 5
\end{array}
$$

$$(x^4 - 3x + 1) \div (x-1) = x^3 + x^2 + x + 4 + \dfrac{5}{x-1}$$

5. $x^8 - 3x^7 + 9x^6 + 1$

$$
\begin{array}{r|rrrrrrrrr}
-3 & 1 & 0 & 0 & 27 & 0 & 0 & 0 & 1 & 3 \\
   &   & -3 & 9 & -27 & 0 & 0 & 0 & 0 & -3 \\
\hline
   & 1 & -3 & 9 & 0 & 0 & 0 & 0 & 1 & 0
\end{array}
$$

$$(x^9 + 27x^6 + x + 3) \div (x+3) = x^8 - 3x^7 + 9x^6 + 1$$

Chapter 2
Factoring and Solving Equations

actoring and solving algebraic equations go together as missions and goals of algebra. An equation set equal to 0 can be solved using the Multiplication Property of Zero—if the portion set equal to 0 is factored. Several techniques discussed in this chapter deal with factoring. Then the chapter finishes with solving equations after using that factoring.

Greatest Common Factors and Factoring Binomials

To factor an expression means to rewrite it as a big multiplication problem—all one term—instead of a string of several terms added (or subtracted) together. For instance, the factored form for $x^4 + 3x^3 - 4x^2 - 12x$ is $x(x + 2)(x - 2)(x + 3)$. The first expression has four terms in it. The factored version has one term—four factors multiplied together. The factored form has many uses in algebra. The factored form is needed in both the numerator and denominator of a fraction, if the fraction is to be reduced. The factored form is needed when solving most nonlinear equations. A factored form usually is more easily evaluated than a nonfactored form. For instance, in the two expressions mentioned earlier, let $x = 4$. When substituting 4 into the first expression, you have to raise it to the fourth power, third power, and so on. You have to work with big numbers. $4^4 + 3(4)^3 - 4(4)^2 - 12(4) = 256 + 3(64) - 4(16) - 48 = 256 + 192 - 64 - 48 = 336$. I don't know about you, but I couldn't do that one in my head. Compare that to putting the 4 into the factored form: $x(x + 2)(x - 2)(x + 3) = 4(4 + 2)(4 - 2)(4 + 3) = 4(6)(2)(7) = 24 \cdot 14 = 336$. Okay, I couldn't do the last multiplication in my head either, but it was still easier!

Greatest Common Factor

Writing an expression in factored form seems almost like undoing distribution. Instead of multiplying every term by the same factor as done using the distributive property, factoring actually *divides* every term by the same factor and leaves the division result behind, inside the parentheses. No remainder occurs in this division; it divides evenly. The first step to factoring out the Greatest Common Factor (GCF) is to determine what the GCF is. In general, a common factor is a multiplier shared by every term in an expression. The GCF is the largest possible factor—containing all of the smaller factors—that divides every term in the expression evenly, leaving no remainders or fractions.

Example Problems

These problems show the answers and solutions.

1. Factor out the GCF of $30x^4y^2 + 12x^3y - 14x^2$.

   answer: $2x^2(15x^2y^2 + 6xy - 7)$

   The GCF is $2x^2$ because each term is divisible evenly by both 2 and x^2. No factor of y is evident in the last term. The question now is $2x^2(\ ???\ ) = 30x^4y^2 + 12x^3y - 14x^2$.

   Divide each term by $2x^2$ and put the result of each division in the parentheses.

   $$\frac{30x^4y^2}{2x^2} = 15x^2y^2, \quad \frac{12x^3y}{2x^2} = 6xy, \quad \frac{-14x^2}{2x^2} = -7$$

   So, $30x^4y^2 + 12x^3y - 14x^2 = 2x^2(15x^2y^2 + 6xy - 7)$. This is completely factored.

2. Factor $-20ab^2 + 30a^2b^3 - 5ab^4$.

   answer: $-5ab^2(4 - 6ab + b^2)$

   The GCF is either $5ab^2$ or $-5ab^2$. Either divides evenly. Having a lead coefficient in the parentheses that is positive is nicer, so $-5ab^2$ is a better choice. Divide each term by $-5ab^2$.

   $$\frac{-20ab^2}{-5ab^2} = 4, \quad \frac{30a^2b^3}{-5ab^2} = -6ab, \quad \frac{-5ab^4}{-5ab^2} = b^2$$

   The GCF multiplies each of these results.

3. Factor $9x^2z(y - 3)^2 + 12x(y - 3)^3 - 3xw(y - 3)$.

   answer: $3x(y-3)\left[3xz(y-3)+4(y-3)^2-w\right]$

   The three terms here have common factors of 3, x, and $(y - 3)$ making the GCF equal to $3x(y - 3)$.

   $$\frac{9x^2z(y-3)^2}{3x(y-3)} = 3xz(y-3), \quad \frac{12x(y-3)^3}{3x(y-3)} = 4(y-3)^2, \quad \frac{-3xw(y-3)}{3x(y-3)} = -w$$

   Since the terms in the brackets won't combine, even if you distribute the $3xz$ or the 4 over the terms in the parentheses, just leave it as is.

4. Factor $a^{1/2}b^{-1/3} + a^{3/2}b^{-4/3}$.

   answer: $a^{1/2}b^{-4/3}(b + a)$

   The two terms have two factors in common. The factors are powers of a and b. To choose the powers correctly, determine which of the two powers of a is smaller. The number $\frac{1}{2}$ is smaller than $\frac{3}{2}$, so the common factor that can be taken out is $a^{1/2}$. The two powers of b are $-\frac{1}{3}$ and $-\frac{4}{3}$. The smaller of those two is $-\frac{4}{3}$, so the common factor of the two preceding terms is $b^{-4/3}$. Putting them all together, the GCF of the two terms is $a^{1/2}b^{-4/3}$.

$$\frac{a^{1/2}b^{-1/3}}{a^{1/2}b^{-4/3}} = a^{1/2-(1/2)}b^{-1/3-(-4/3)} = a^0b^1, \frac{a^{3/2}b^{-4/3}}{a^{1/2}b^{-4/3}} = a^{3/2-1/2}b^{-4/3-(-4/3)} = a^1b^0$$

When dividing values with the same base, you subtract the exponents. So,

$$a^{1/2}b^{-1/3} + a^{3/2}b^{-4/3} = a^{1/2}b^{-4/3}(a^0b^1 + a^1b^0) = a^{1/2}b^{-4/3}(b+a)$$

Work Problems

Use these problems to give yourself additional practice.

1. Factor $6m^2n^3 + 9mn^4 - 12m^3n^2$.

2. Factor $121a(b-1)^2 - 11a^2(b-1)$.

3. Factor $12x^2y^3z + 16x^3y^2z^2 - 32xy^4z^3$.

4. Factor $10r^{1/2}t^{3/2} - 5r^{3/2}t^{-1/2}$.

5. Factor $-4w^{-1} - 8w^{-2} - 16w^{-3}$.

Worked Solutions

1. **$3mn^2(2mn + 3n^2 - 4m^2)$**   The GCF is $3mn^2$; divide each term by the GCF.
$\frac{6m^2n^3}{3mn^2} = 2mn, \frac{9mn^4}{3mn^2} = 3n^2, \frac{-12m^3n^2}{3mn^2} = -4m^2$. Writing the GCF as the multiplier of
the division results, the factored form is $3mn^2(2mn + 3n^2 - 4m^2)$.

2. **$11a(b-1)[11(b-1) - a]$**   The GCF is $11a(b-1)$. Divide each term by the GCF.
$\frac{121a(b-1)^2}{11a(b-1)} = 11(b-1), \frac{-11a^2(b-1)}{11a(b-1)} = -a$. Multiplying the GCF times the two results,
you get $11a(b-1)[11(b-1) - a]$. Even if you distribute the 11 in the brackets, nothing
will combine, so you can just leave it that way.

3. **$4xy^2z(3xy + 4x^2z - 8y^2z^2)$**   Dividing each term by the GCF, $4xy^2z$, yields
$\frac{12x^2y^3z}{4xy^2z} = 3xy, \frac{16x^3y^2z^2}{4xy^2z} = 4x^2z, \frac{-32xy^4z^3}{4xy^2z} = -8y^2z^2$. The answer is the product of
that GCF and the sum of the division results. The three terms in the parentheses have
nothing in common (all three at the same time).

4. **$5r^{1/2}t^{-1/2}(2t^2 - r)$**   The GCF contains factors whose exponents are the smallest of the
exponents in the terms. The factor r has exponents $\frac{1}{2}$ and $\frac{3}{2}$; the smaller is the $\frac{1}{2}$. The
factor t has exponents of $\frac{3}{2}$ and $-\frac{1}{2}$. The smaller of those is $-\frac{1}{2}$.

$$\frac{10r^{1/2}t^{3/2}}{5r^{1/2}t^{-1/2}} = 2t^{3/2-(-1/2)} = 2t^2, \frac{-5r^{3/2}t^{-1/2}}{5r^{1/2}t^{-1/2}} = -r^{3/2-1/2} = -r^1$$

Writing the GCF times the two division results yields $10r^{1/2}t^{3/2} - 5r^{3/2}t^{-1/2} = 5r^{1/2}t^{-1/2}(2t^2 - r)$.

5. **$-4w^{-3}(w^2 + 2w + 4)$**   The GCF is $-4w^{-3}$. The -4 is used instead of $+4$ so there won't be three negative signs inside the parentheses. The -3 exponent of w is used because it's the smallest of the three exponents. Dividing each term by the GCF, yields $\dfrac{-4w^{-1}}{-4w^{-3}} = w^{-1-(-3)} = w^2$, $\dfrac{-8w^{-2}}{-4w^{-3}} = 2w^{-2-(-3)} = 2w^1$, $\dfrac{-16w^{-3}}{-4w^{-3}} = 4$. The factored form has the GCF multiplying the results of these three divisions.

Factoring Binomials

A binomial is an expression with two terms separated by either addition or subtraction. The goal is to make the binomial all one term instead of two separate terms—with everything multiplied together. This is accomplished by factoring the two terms. You can use four basic methods to factor a binomial. If none of these methods works, the expression is considered to be **prime;** it cannot be factored.

The rules or patterns to use when doing the factoring are as follows:

Rule 1.   Factoring out the Greatest Common Factor

$$ab + ac = a(b + c)$$

The common factor here is a; it's written as a multiplier of what is left after dividing it out of the terms.

Rule 2.   Factoring using the pattern for the *difference of squares*

$$a^2 - b^2 = (a - b)(a + b)$$

Using one of the rules for special products of binomials (discussed earlier in the chapter), this is how the difference between two squares can be written as a product.

Rule 3.   Factoring using the pattern for the *difference of cubes*

$$a^3 - b^3 = (a - b)(a^2 + ab + b^2)$$

This rule and the next use a similar pattern. There are two factors—a binomial and a trinomial. The binomial contains the two cube roots of the terms. The trinomial contains the squares of those roots in the first and third positions and then the opposite of the product of those two cube roots as the middle term.

Rule 4.   Factoring using the pattern for the *sum of cubes*

$$a^3 + b^3 = (a + b)(a^2 - ab + b^2)$$

Just like the preceding rule, two factors exist—a binomial and a trinomial. The binomial contains the two cube roots of the terms. The trinomial contains the squares of those roots in the first and third positions and then the opposite of the product of those two roots as the middle term.

The challenge will be in determining which factoring method to use. If you recognize that both terms are perfect squares and they are subtracted, then Rule 2 makes sense. If both terms are perfect cubes, then Rule 3 or Rule 4 will work. If both terms have one or more factors in common, then use Rule 1. Sometimes, you get to use more than one rule to complete the job. It's best to check for a GCF first to make the values in the terms smaller and other patterns more apparent.

Example Problems

These problems show the answers and solutions.

1. Factor $18a^3b^2 - 27a^4$.

    answer: $9a^3(2b^2 - 3a)$

    Neither term is a perfect square nor a perfect cube. They do have a common factor, though, which is $9a^3$.

2. Factor $25y^2 - 16$.

    answer: $(5y + 4)(5y - 4)$

    The two terms are each perfect squares, and the expression fits Rule 2, where you have the difference between two squares. The answer is the product of the sum and difference of the roots of the two numbers.

3. Factor $z^6 - 144$.

    answer: $(z^3 + 12)(z^3 - 12)$

    The two terms are each perfect squares, and they're being subtracted, so this involves Rule 2 again. The term is $z^6 = (z^3)^2$, so the factored form has the sum and difference of the two roots of the squares. Even though a z^3 is in each factor, which is a cube, the 12 is not a perfect cube, so the result can't be factored further.

4. Factor $m^3 - 125$.

    answer: $(m - 5)(m^2 + 5m + 25)$

    This is the difference between two perfect cubes, so Rule 3 is used here. The rule for doing this factorization involves finding the two cube roots. If the two cubes are a^3 and b^3, then their roots are a and b, and those two roots are written in the first part of the factorization, $(a - b)$. Then a trinomial is built using the squares of those two roots and the *opposite of the product* of those same two roots. In this problem, m is the cube root of m^3, and -5 is the cube root of -125. The square of m is m^2, and the square of -5 is 25. The opposite of the product of those two cube roots is the opposite of $-5m$, which is $+5m$. Now, put everything into the pattern, $m^3 - 125 = (m - 5)(m^2 + 5m + 25)$.

5. Factor $27 + 8a^3$.

    answer: $(3 + 2a)(9 - 6a + 4a^2)$

    The two terms are each perfect cubes and, because they're added, Rule 4 is used. $27 = (3)^3$, so 3 is the cube root of 27, and $8a^3 = (2a)^3$, so $2a$ is the cube root of $8a^3$. Using the same pattern as in the preceding example, the two cube roots are written in the binomial, and their squares are the first and last terms in the trinomial. The opposite of the product of these two cube roots is $-6a$, so it's written as the middle term in the trinomial.

    $$27 + 8a^3 = (3 + 2a)(9 - 6a + 4a^2)$$

6. Factor $x^3r^2 - 4x^3t^2$.

    answer: $x^3(r + 2t)(r - 2t)$

    Even though squares and cubes are in both terms, this expression doesn't fit any of the patterns for factoring differences of cubes or squares. A common factor exists in the two terms, though—x^3. Factoring that out, you get $x^3(r^2 - 4t^2)$, which has something nice. The expression in the parentheses can be factored using the difference of squares. So, factoring further yields $x^3(r^2 - 4t^2) = x^3(r + 2t)(r - 2t)$.

Work Problems

Use these problems to give yourself additional practice.

Factor each expression.

1. $16a^2b^3 - 40a^3b^6$

2. $16a^2b^4 - 49$

3. $1 - 64x^3$

4. $343 + y^3z^6$

5. $x^5 + x^2$

Worked Solutions

1. **$8a^2b^3(2 - 5ab^3)$**   This expression has a common factor of $8a^2b^3$. The result of the division of the two terms by that the GCF is written in the parentheses.

2. **$(4ab^2 + 7)(4ab^2 - 7)$**   The two terms are both perfect squares. Rule 2 is used, and the factored form has the product of the sum and difference of the two roots.

3. **$(1 - 4x)(1 + 4x + 16x^2)$**   The two terms are perfect cubes. This uses Rule 3, in which the result is a binomial containing the cube roots of the two terms times a trinomial containing the squares of the two roots and the opposite of their product.

4. **$(7 + yz^2)(49 - 7yz^2 + y^2z^4)$**   This is the sum of two cubes, which uses Rule 4.

5. **$x^2(x + 1)(x^2 - x + 1)$**   This problem required applying two different rules. First, the GCF is factored out, leaving the sum of two cubes. Then the sum of the cubes is factored using Rule 4: $x^5 + x^2 = x^2(x^3 + 1) = x^2(x + 1)(x^2 - x + 1)$.

Factoring Trinomials

Trinomials can be factored in two ways—either by finding a Greatest Common Factor or by un-FOILing (or both). Factoring out the GCF is discussed earlier in this chapter under "Greatest Common Factor." The other option, un-FOILing, involves determining which two binomials were multiplied together to get the trinomial. Sometimes, both methods will be used. A common factor *and* two binomials whose product is the trinomial can exist in the same problem. In those cases, it is best to take out the common factor first, to make the numbers smaller.

If you need to review multiplying binomials together using the FOIL method, refer to the section "The FOIL Method" in Chapter 1. Just as a quick review/reference, consider multiplying the two binomials $(a + b)(c + d)$. When using FOIL, you add the products of the First terms, Outer terms, Inner terms, and Last terms. So, $(a + b)(c + d) = a \cdot c + a \cdot d + b \cdot c + b \cdot d$.

Getting back to factoring a trinomial, look at the general trinomial $ax^2 \pm bx \pm c$. The $\pm$ means that either an add or a subtract can be in that position. Using this trinomial, the procedure for factoring the trinomial by un-FOILing is as follows:

First, find two terms whose product is the first term in the trinomial, ax^2. These two terms will be in the First positions in the two binomials of the factorization. If the term ax^2 is negative, it's a good idea to factor -1 out of each term so that you can start with a positive term.

Second, find two terms whose product is the last term in the trinomial, c. These two terms will be in the Last positions in the binomials.

Next, if c is **positive**, arrange the terms determined previously so that the Outer and Inner products will have a **sum** equal to bx, the middle term in the trinomial.

If c is **negative**, arrange the terms determined previously so that the Outer and Inner products will have a **difference** of bx, the middle term.

Next comes choosing which terms to use and arranging them in the binomials. The Outer and Inner products will either be added or subtracted, depending on their signs.

Example Problems

These problems show the answers and solutions.

1. Factor $x^2 - 2x - 15$.

    answer: $(x - 5)(x + 3)$

    The two terms whose product is x^2 are x and x. You have two choices for the product of 15: $1 \cdot 15$ and $3 \cdot 5$. The possible arrangements for these terms are $(x \quad 1)(x \quad 15)$ or $(x \quad 3)(x \quad 5)$. The Outer and Inner products need to have a difference of $2x$, the middle term. That makes the choice of factors $(x \quad 3)(x \quad 5)$. There will have to be a $+$ in one of the binomials and a $-$ in the other, so that the Last product will be negative. The arrangement that gives the correct product with the correct sign on the middle term is $(x - 5)(x + 3)$.

2. Factor $-3x^2 + 10x - 8$.

    answer: $-(3x - 4)(x - 2)$

    Here, it's a good idea to factor out -1 to make the lead term positive: $-3x^2 + 10x - 8 = -1(3x^2 - 10x + 8)$. Now concentrate on the trinomial inside the parentheses. The two terms whose product is $3x^2$ are $3x$ and x. You have two different choices for the product of 8: $1 \cdot 8$ and $2 \cdot 4$. The possible arrangements for these terms are $(3x \quad 1)(x \quad 8)$, $(3x \quad 8)(x \quad 1)$, $(3x \quad 2)(x \quad 4)$, or $(3x \quad 4)(x \quad 2)$. Because the 8 is positive, the Outer and Inner products need to have a sum of $10x$, the middle term. The choice that accomplishes this is $(3x \quad 4)(x \quad 2)$. The Outer product is $6x$, and the Inner product is $4x$. Their sum is $10x$. To make the signs of the product come out correctly, put a $-$ in each binomial and a $-$ in front for the -1 that was factored out at the beginning: $-(3x - 4)(x - 2)$.

3. Factor $30y^3 - 33y^2 - 18y$.

    answer: $3y(2y - 3)(5y + 2)$

    The first step is to factor out the GCF and then look at the resulting trinomial to see whether it can be un-FOILed. The GCF is $3y$, and the factored version is $3y(10y^2 - 11y - 6)$. The first term in the trinomial can have factors of $10y$ and y or $5y$ and $2y$. The last term can have factors of 1 and 6 or of 2 and 3. This time, the factors have to be arranged so that there will be a difference of $11y$ between the Outer and Inner products. This can be accomplished with the product $(2y\ 3)(5y\ 2)$. The Outer product is $4y$, and the Inner product is $15y$. Their difference is $11y$. A $+$ and $-$ sign will have to be assigned so that the difference is negative. This means that you want the $15y$ to be negative. The arrangement that works is $(2y - 3)(5y + 2)$. Put the $3y$ in front to complete the factorization: $30y^3 - 33y^2 - 18y = 3y(2y - 3)(5y + 2)$.

Work Problems

Use these problems to give yourself additional practice.

1. Factor $3n^2 + 7n + 2$.

2. Factor $5x^2 + 12x - 9$.

3. Factor $-18w^2 + 3w + 10$.

4. Factor $32y^2 - 36y + 4$.

5. Factor $6z^4 - 16z^3 - 6z^2$.

Worked Solutions

1. **$(3n + 1)(n + 2)$**   The first term has factors of $3n$ and n. The last term has factors of 1 and 2. The factors need to be arranged so that the sum of the Outer and Inner products is $7n$. This is accomplished with $(3n\ 1)(n\ 2)$. Since the middle term is positive, both of the signs in the binomials are positive.

2. **$(5x - 3)(x + 3)$**   The first term has factors of $5x$ and x. The last term has factors of 1 and 9, or 3 and 3. The last term is negative, so the Outer and Inner products have to have a difference of $12x$. Arranging the factors $(5x\ 3)(x\ 3)$ gives an Outer product of $15x$ and an Inner product of $3x$. Their difference is $12x$. The signs are placed so that the middle term is positive.

3. **$-(6w - 5)(3w + 2)$**   The first thing to do is to factor out a -1 so that the first term in the trinomial will be positive. $-18w^2 + 3w + 10 = -1(18w^2 - 3w - 10)$. The factors of 18 are 1 and 18, 2 and 9, or 3 and 6. The factors of 10 are 1 and 10, or 2 and 5. The correct ones are chosen so that the Outer and Inner products will have a difference of 3. This is accomplished with $(6w\ 5)(3w\ 2)$. The middle term is negative, so the Outer product has to be negative, which can be done if the 2 is negative. The final factorization, including the -1 is $-1(6w - 5)(3w + 2)$.

4. **$4(y - 1)(8y - 1)$**   The first thing to do is to factor out a 4 from each term. This makes all of the coefficients smaller and more manageable. $32y^2 - 36y + 4 = 4(8y^2 - 9y + 1)$. The trinomial can be factored using $1y$ and $8y$ for the factors of the first term. The sum of the Outer and Inner products is $9y$, and both signs are negative: $8y^2 - 9y + 1 = (8y - 1)(y - 1)$. Multiply this product by 4 to complete the answer.

5. **$2z^2(z - 3)(3z + 1)$**   First factor the GCF of $2z^2$ from each term: $6z^4 - 16z^3 - 6z^2 = 2z^2(3z^2 - 8z - 3)$. The factors of the first term are z and $3z$. The factors of the last term of the trinomial are 1 and 3. The difference between the Outer and Inner products has to be $8z$, and this is done by arranging the products as $(z \quad 3)(3z \quad 1)$. The middle term is negative, so the Inner product has to be negative, which is accomplished by making the 3 negative: $3z^2 - 8z - 3 = (z - 3)(3z + 1)$. Put the GCF in front to get the final answer.

Factoring Other Polynomials

Two processes considered in this section are

- ❑ Factoring quadratic-like expressions
- ❑ Factoring by grouping

Quadratic-Like Expressions

A quadratic-like expression is one of the form $ax^{2n} + bx^n + c$ where there is a trinomial with one term raised to an even power, a second term raised to half that even power, and a constant. Some examples are $y^4 + 6y^2 + 5$, $6a^6 - a^3 - 2$, and $9z^{-2} - 30z^{-1} + 25$. These can be factored using the same un-FOIL process that is used to factor quadratics such as $y^2 - 9y - 10$. The same First, Outer, Inner, and Last pattern exists. Go back to products of binomials and/or factoring trinomials for more information on this.

Example Problems

These problems show the answers and solutions.

1. Factor $x^4 + 2x^2 - 35$.

    answer: $(x^2 - 5)(x^2 + 7)$

    The First term can be written as $x^2 \cdot x^2$. The Last term can be written as $5 \cdot 7$. Because the last term is negative, you're looking for the *difference* of the Outer and Inner products to be equal to the middle term, the $2x^2$. The middle term is positive, so the sign of the 7 will be $+$.

2. Factor $12m^{-6} - 13m^{-3} + 3$.

    answer: $(3m^{-3} - 1)(4m^{-3} - 3)$

    The First term can be written as either $1m^{-3} \cdot 12m^{-3}$, $2m^{-3} \cdot 6m^{-3}$, or $3m^{-3} \cdot 4m^{-3}$, and the Last term can be written as $3 \cdot 1$. The last term is positive, so you're looking for the *sum* of the Outer and Inner products to be equal to the middle term, $13m^{-3}$. The middle term is negative, so there will be two $-$ signs in the binomials. The combination that gives the $13m^{-3}$ in the middle term uses the $3m^{-3}$ and $4m^{-3}$ with the $3m^{-3}$ and the 3 both in the Outer position.

3. Factor $a^8 - 18a^4 - 175$.

   answer: $(a^2 - 5)(a^2 + 5)(a^4 + 7)$

   The First term can be written as $a^4 \cdot a^4$. The Last term can be written as $175 \cdot 1$, $5 \cdot 35$, or $7 \cdot 25$. To get a difference of $-18a^4$ in the middle, the combination $(a^4 \quad 25)(a^4 \quad 7)$ is used, with the 25 being negative and the 7 positive. This factored form, $(a^4 - 25)(a^4 + 7)$, has a factor that can be factored. The first factor is the difference of two squares, which is factored into the sum and difference of the roots. The a^4 can be written as $a^2 \cdot a^2$ and the 25 as $5 \cdot 5$.

Factoring by Grouping

Factoring by grouping is a method that is usually done on expressions with four, six, eight, or more terms. Typically, no common factor exists for all of the terms in the expression, but common factors do exist in pairs of the terms. The common factor is identified for each pair or grouping, the factoring is done, and then the new expression, with new terms, is checked to see whether there is a common factor throughout. If the factoring of separate groups doesn't result in a common factor in the new terms, try rearranging the terms of the original problem into different orders and different groups. It could be that the expression just can't be factored, but different arrangements should be tried first.

Example Problems

These problems show the answers and solutions.

1. Factor by grouping $a^2x^2 + 5a^2 + 6x^2 + 30$.

   answer: $(a^2 + 6)(x^2 + 5)$

   As you can see, no factor is common to all four terms at once. The first two terms have a common factor of a^2, and the last two terms have a common factor of 6. Factor that a^2 out of the first two terms and the 6 out of the last two: $a^2x^2 + 5a^2 + 6x^2 + 30 = a^2(x^2 + 5) + 6(x^2 + 5)$. There are now two terms instead of four, and the two terms have a common factor of $x^2 + 5$. If this grouping factoring had not resulted in this nice situation, you would have had to try another arrangement of the terms. This type of factoring only works when the expression cooperates like this. The next step is to factor the $x^2 + 5$ out of each term. Factoring that out of the first term, you are left with a^2, and factoring that out of the second term leaves you 6. The final result is $(a^2 + 6)(x^2 + 5)$.

2. Factor by grouping $x^2y^2 - x^2 - 9y^2 + 9$.

   answer: $(y + 1)(y - 1)(x + 3)(x - 3)$

   Again, you have no factor common to all four terms at once. The first two terms have a common factor of x^2, and the last two terms have a common factor of 9 or -9. Using the x^2 and 9 and putting these two factorizations together yields $x^2y^2 - x^2 - 9y^2 + 9 = x^2(y^2 - 1) + 9(-y^2 + 1)$. This time, the two new terms don't have a common factor. The two factors in the parentheses aren't exactly alike. They are opposite in sign, however, so the factorization of those last two terms should be revisited. Factor out a -9 instead of 9. $x^2y^2 - x^2 - 9y^2 + 9 = x^2(y^2 - 1) - 9(y^2 - 1)$. Now the two terms have the common factor of $y^2 - 1$. Factor that out to get $x^2(y^2 - 1) - 9(y^2 - 1) = (y^2 - 1)(x^2 - 9)$. Both of those factors can be factored because they are the difference of perfect squares: $(y^2 - 1)(x^2 - 9) = (y + 1)(y - 1)(x + 3)(x - 3)$.

3. Factor $2ax^2 + 6x^2 - ax - 3x - 3a - 9$ using grouping.

answer: $(a + 3)(2x - 3)(x + 1)$

No factor is common to all six terms. The first two terms have a common factor of $2x^2$; the second two terms have a common factor of $-x$; and the last two terms have a common factor of -3. Do these three factorizations to get $2ax^2 + 6x^2 - ax - 3x - 3a - 9 = 2x^2(a + 3) - x(a + 3) - 3(a + 3)$. Now there are three terms instead of six, and each has the common factor of $(a + 3)$. Factoring that out, $2x^2(a + 3) - x(a + 3) - 3(a + 3) = (a + 3)(2x^2 - x - 3)$. And, just when you think you're finished, you notice that the trinomial in this result can be factored. Using un-FOIL, the final factorization is $(a + 3)(2x^2 - x - 3) = (a + 3)(2x - 3)(x + 1)$.

The preceding example could have been factored, by grouping, using groups of three instead of two. With some rearranging and getting the three terms with a in them first, the last three terms then have a common factor of 3. It was just as easy to do it in pairs of factors. There's usually more than one way to group and factor these problems, so you can't go wrong, as long as you follow the rules.

Work Problems

Use these problems to give yourself additional practice.

1. Factor $c^{4/3} + 2c^{2/3} - 48$.

2. Factor $x^8 + 9x^4 + 8$.

3. Factor $ac - 3ab + 2bc - 6b^2$.

4. Factor $48xy + 30x - 40y - 25$.

5. Factor $9a^3x^2 + 9x^2 - a^3 - 1$.

Worked Solutions

1. $\left(c^{2/3} - 6\right)\left(c^{2/3} + 8\right)$ This is a quadratic-like situation. The exponent of the first factor is twice that of the second factor. Using un-FOIL to factor the trinomial, the First term is $c^{2/3} \cdot c^{2/3}$. The Last term can be $1 \cdot 48$, $2 \cdot 24$, $3 \cdot 16$, $4 \cdot 12$, or $6 \cdot 8$, but the one to use is $6 \cdot 8$ because the difference between those factors is 2. The signs in $\left(c^{2/3} \quad 6\right)\left(c^{2/3} \quad 8\right)$ will be $+$ and $-$, with the positive sign on the 8, because the middle term is positive.

2. $(x^4 + 1)(x^4 + 8)$ This is another quadratic-like problem. The sum of the Outer and Inner products is $9x^4$, so both binomials will have $+$ signs.

3. $(c - 3b)(a + 2b)$ This problem is factored using grouping. The first two terms have a common factor of a, and the second two have a common factor of $2b$. Factoring those out, $ac - 3ab + 2bc - 6b^2 = a(c - 3b) + 2b(c - 3b)$. Now the two terms have a common factor of $c - 3b$. When that's factored out, it leaves the a and $2b$ in the second binomial.

4. **(8y + 5)(6x − 5)**   This is factored using grouping. The first two terms have a common factor of $6x$, and the second two terms have a common factor of -5: $48xy + 30x - 40y - 25 = 6x(8y + 5) - 5(8y + 5)$. Now the factor of $8y + 5$ can be factored out of the two terms.

5. **(a + 1)(a² − a + 1)(3x + 1)(3x − 1)**   This is first factored using grouping, $9a^3x^2 + 9x^2 - a^3 - 1 = 9x^2(a^3 + 1) - 1(a^3 + 1) = (a^3 + 1)(9x^2 - 1)$. Then the two binomials are factored. The first is the sum of two cubes, and the second is the difference between two squares: $(a^3 + 1)(9x^2 - 1) = (a + 1)(a^2 - a + 1)(3x + 1)(3x - 1)$.

Solving Quadratic Equations

A quadratic equation is a statement with a variable in it that's raised to the second power, or squared. This squared term is the highest power in the expression. If the equation is quadratic, then there can be two solutions, one solution, or no solutions to the equation. The two solutions are two distinct (different) numbers. One solution means that an answer has been repeated—the same thing appears twice. And "no solution" means that there's no real number that satisfies the equation. To solve a quadratic equation, you can use four possible methods:

1. *BGBG: By Guess or By Golly*

   This method isn't very efficient, but sometimes solutions seem obvious, and there's no harm in *knowing* the solution!

2. *Factoring*

   This is probably the most efficient way of solving a quadratic equation—as long as the solutions are rational numbers, which means the trinomial is factorable. The factored form and the Multiplication Property of Zero join together to make this a preferred method.

3. *Quadratic Formula*

   The quadratic formula *always* works. Whether an equation factors or not, you can always use the quadratic formula to get the answers. Why isn't it *always* used then? The quadratic formula can be rather messy and cumbersome, and errors can be made more easily using this than in factoring. This formula is a second resort.

4. *Completing the Square*

   This method should *never* be used to solve a quadratic equation, but students are asked to do just that. Why is that so? Completing the square is an important technique used to rewrite equations of circles, parabolas, hyperbolas, and ellipses in their standard form—a more useful form. By teaching completing the square to algebra students, they get practice for later. And, this method does give the answers to the quadratic equation, as well. Go to Chapter 8 if you want to see completing the square in action.

Solving Quadratic Equations by Factoring

When solving a quadratic equation by factoring, you're taking advantage of the Multiplication Property of Zero. See Chapter 1, "The Basics," for more on that property. The Multiplication Property of Zero says that if the product of values is equal to zero, then one of the values must be zero. This works with factoring the quadratic, because when the factored expression is set equal to zero, then each of the factors can be set equal to zero, independently, to see what value of the variable would make it so.

Example Problems

These problems show the answers and solutions.

1. Solve for x by factoring $x^2 + 6x - 7 = 0$.

    answer: $x = -7$ or $x = 1$

    The trinomial factors, giving $(x + 7)(x - 1) = 0$. The Multiplication Property of Zero says that one or the other of the factors must be equal to zero, so $x + 7 = 0$ or $x - 1 = 0$. The two solutions to this quadratic equation are $x = -7$ or $x = 1$.

2. Solve for y by factoring $3y^2 - 11y = 4$.

    answer: $y = -\frac{1}{3}$ or $y = 4$

    This equation must first be set equal to 0. Subtract 4 from each side. Now the trinomial can be factored.

    $$3y^2 - 11y - 4 = (3y + 1)(y - 4) = 0$$

    One or the other of the factors must equal 0. If $3y + 1 = 0$, then $3y = -1$, and $y = -\frac{1}{3}$. If $y - 4 = 0$, then $y = 4$.

3. Solve for z by factoring $11z = z^2$.

    answer: $z = 0$ or $z = 11$

    A common error in solving this equation is to divide each side by z. As a rule, don't divide by the variable—you'll lose a solution. It's okay to divide each side by a constant number but not a variable. Set the equation equal to zero by subtracting $11z$ from each side.

    $$\begin{aligned} 11z &= z^2 \\ 0 &= z^2 - 11z \end{aligned}$$

    Now factor out the z: $0 = z(z - 11)$. One of the factors must be equal to 0, $z = 0$ or $z - 11 = 0$.

    As you see, one of the solutions just pops up. If $z - 11 = 0$, then $z = 11$, so the two solutions are $z = 0$ or $z = 11$.

4. Solve for w in $w^2 = 25$.

    answer: $w = 5$ or $w = -5$

    This can actually be done in one of two ways, by factoring or by taking the square root of each side. Using factoring, first get the equation to equal to 0, then factor, and then solve for the solutions.

    $$\begin{aligned} w^2 = 25 &= 0 \\ (w - 5)(w + 5) &= 0 \end{aligned}$$

If $w - 5 = 0$, then $w = 5$. If $w + 5 = 0$, then $w = -5$. The two solutions are $w = 5$ or $w = -5$.

Taking the square root of each side, yields

$$w^2 = 25$$
$$\sqrt{w^2} = \sqrt{25}$$

$\sqrt{w^2} = |w|$ because w could be positive or negative, and both of these have to be considered when taking the square root of both sides. When doing this, write

$$\sqrt{w^2} = \sqrt{25}$$
$$w = \pm 5$$

Just as with the factoring, the solutions are $w = 5$ or $w = -5$.

5. Solve for t by factoring $4t^2 + 12t + 9 = 0$.

    answer: $t = -\dfrac{3}{2}$

    Factor the trinomial $4t^2 + 12t + 9 = (2t + 3)(2t + 3) = 0$. Setting the first factor equal to 0, yields

    $$2t + 3 = 0$$
    $$2t = -3$$
    $$t = -\frac{3}{2}$$

    The same thing happens with the second factor. This is a case of a quadratic equation having a double root—two roots that are exactly the same. This happens when the quadratic trinomial is a perfect square trinomial.

Work Problems

Use these problems to give yourself additional practice. Solve each quadratic equation by factoring.

1. $x^2 - 14x + 45 = 0$

2. $6y^2 - 7y = 3$

3. $m^2 - 20m + 100 = 0$

4. $7p^2 = 63$

5. $16 - 40z + 25z^2 = 0$

Worked Solutions

1. **$x = 5, x = 9$**   First factor the trinomial, $x^2 - 14x + 45 = (x - 5)(x - 9) = 0$. Set the two factors equal to zero and solve the equations: $x - 5 = 0$, $x = 5$; $x - 9 = 0$, $x = 9$.

2. $y = -\frac{1}{3}$, $y = \frac{3}{2}$ First rewrite the equation so it's set equal to 0: $6y^2 - 7y - 3 = 0$. Next factor the trinomial and set the two factors equal to zero:

$$6y^2 - 7y - 3 = (3y + 1)(2y - 3) = 0$$
$$3y + 1 = 0, \ y = -\frac{1}{3}; \ 2y - 3 = 0, \ y = \frac{3}{2}$$

3. $m = 10$ This equation has a double root. When it's factored, the same factor appears twice: $m^2 - 20m + 100 = (m - 10)(m - 10) = 0$. Setting that factor equal to zero, yields $m - 10 = 0$, $m = 10$.

4. $p = 3$, $p = -3$ First divide each side by 7, and then rewrite the equation so it's set equal to 0: $7p^2 = 63$, $p^2 = 9$, $p^2 - 9 = 0$. Factor the binomial and set the factors equal to 0:

$$p^2 - 9 = (p - 3)(p + 3) = 0$$
$$p - 3 = 0, \ p = 3; \ p + 3 = 0, \ p = -3$$

5. $z = \frac{4}{5}$ This equation also has a double root. The two factors are the same:
$16 - 40z + 25z^2 = (4 - 5z)(4 - 5z) = 0$; when $4 - 5z = 0$, $z = \frac{4}{5}$.

Solving Quadratic Equations with the Quadratic Formula

The quadratic formula is just that. It's a formula used to solve for the solutions of a quadratic equation. Many quadratic equations can't be factored, because not all solutions are nice integers or fractions. When the solutions are irrational numbers, with radicals, or if the factoring isn't easy to do, then the quadratic formula is the way to go. In the next section, on completing the square, the quadratic formula is developed—you'll see where it came from. For now, the formula is given, and its use is demonstrated.

The Quadratic Formula

Given the standard form of a quadratic equation, $ax^2 + bx + c = 0$, then the solutions of the equation are

$$x = \frac{-b \pm \sqrt{b^2 - 4ac}}{2a}$$

Example Problems

These problems show answers and solutions.

1. Solve $x^2 - 8x + 7 = 0$ using the quadratic formula.

    answer: $x = 7$ or $x = 1$

    The trinomial can be easily factored, but the quadratic formula works for factorable equations, also. From the standard form of the quadratic equation, in this equation $a = 1$, $b = -8$, and $c = 7$.

Putting those values into the quadratic formula yields

$$x = \frac{8 \pm \sqrt{(-8)^2 - 4(1)(7)}}{2(1)}$$

Notice that, since $b = -8$, the first term in the fraction is the opposite of -8, or $+8$.

$$x = \frac{8 \pm \sqrt{64 - 28}}{2} = \frac{8 \pm \sqrt{36}}{2} = \frac{8 \pm 6}{2}$$

Breaking this up into two fractions, $x = \frac{8 + 6}{2}$ or $x = \frac{8 - 6}{2}$, so $x = \frac{14}{2} = 7$ or $x = \frac{2}{2} = 1$.

When the solutions are integers like this, you can check your work by factoring.

2. Use the quadratic formula to solve $2y^2 + 4y - 9 = 0$.

    answer: $y = \frac{-2 + \sqrt{22}}{2}$ or $y = \frac{-2 - \sqrt{22}}{2}$

    The values of the coefficients are $a = 2$, $b = 4$, and $c = -9$.

    Substituting these into the formula,

    $$x = \frac{-4 \pm \sqrt{4^2 - 4(2)(-9)}}{2(2)} = \frac{-4 \pm \sqrt{16 - (-72)}}{4} = \frac{-4 \pm \sqrt{88}}{4}$$

    The radical can be simplified and the fraction reduced.

    $$\frac{-4 \pm \sqrt{88}}{4} = \frac{-4 \pm \sqrt{4}\sqrt{22}}{4} = \frac{-4 \pm 2\sqrt{22}}{4} = \frac{2(-2 \pm \sqrt{22})}{4_2} = \frac{-2 \pm \sqrt{22}}{2}$$

    The fraction is then broken up into the two answers, $x = \frac{-2 + \sqrt{22}}{2}$ or $x = \frac{-2 - \sqrt{22}}{2}$.

3. Use the quadratic formula to solve $3z^2 - 2z + 5 = 0$.

    answer: There's no real answer.

    The coefficients are $a = 3$, $b = -2$, and $c = 5$.

    Substitute into the formula,

    $$z = \frac{2 \pm \sqrt{(-2)^2 - 4(3)(5)}}{2(3)} = \frac{2 \pm \sqrt{4 - 60}}{6} = \frac{2 \pm \sqrt{-56}}{6}$$

You can stop right there. You can't take the square root of a negative number and get a real answer, so this equation has no real solution. Refer to "Complex Numbers" in Chapter 7 for information on how to deal with this situation.

Work Problems

Use these problems to give yourself additional practice. Use the quadratic formula to solve each.

1. $x^2 - 7x - 18 = 0$

2. $10y^2 - 13y = 3$

3. $4w^2 - 35w + 24 = 0$

4. $4x^2 - 25 = 0$

5. $z^2 + 10z - 8 = 0$

Worked Solutions

1. **$x = 9$, $x = -2$**   The coefficients are $a = 1$, $b = -7$, and $c = -18$.

   Substituting into the formula,

   $$x = \frac{7 \pm \sqrt{(-7)^2 - 4(1)(-18)}}{2(1)} = \frac{7 \pm \sqrt{49 - (-72)}}{2} = \frac{7 \pm \sqrt{121}}{2} = \frac{7 \pm 11}{2}$$

   Breaking up the fraction yields $x = \frac{7 + 11}{2} = \frac{18}{2} = 9$ or $x = \frac{7 - 11}{2} = \frac{-4}{2} = -2$

2. **$y = \frac{3}{2}$, $y = -\frac{1}{5}$**   First, write the equation in standard form by subtracting 3 from each side: $10y^2 - 13y - 3 = 0$

   The coefficients are $a = 10$, $b = -13$, and $c = -3$.

   Substituting into the formula,

   $$y = \frac{13 \pm \sqrt{(-13)^2 - 4(10)(-3)}}{2(10)} = \frac{13 \pm \sqrt{169 - (-120)}}{20} = \frac{13 \pm \sqrt{289}}{20} = \frac{13 \pm 17}{20}$$

   Breaking up the fraction yields $y = \frac{13 + 17}{20} = \frac{30}{20} = \frac{3}{2}$ or $y = \frac{13 - 17}{20} = \frac{-4}{20} = -\frac{1}{5}$

3. **$w = 8$, $w = \frac{3}{4}$**   The coefficients are $a = 4$, $b = -35$, and $c = 24$.

   Substituting into the formula,

   $$w = \frac{35 \pm \sqrt{(-35)^2 - 4(4)(24)}}{2(4)} = \frac{35 \pm \sqrt{1225 - 384}}{8} = \frac{35 \pm \sqrt{841}}{8} = \frac{35 \pm 29}{8}$$

   Breaking up the fraction yields $w = \frac{35 + 29}{8} = \frac{64}{8} = 8$ or $w = \frac{35 - 29}{8} = \frac{6}{8} = \frac{3}{4}$

4. $x = \frac{5}{2}$, $x = -\frac{5}{2}$ There are only two terms in this quadratic. The middle term is missing, so the coefficient of b is equal to 0.

The coefficients are $a = 4$, $b = 0$, and $c = -25$.

Substituting into the formula,

$$x = \frac{0 \pm \sqrt{0^2 - 4(4)(-25)}}{2(4)} = \frac{\pm\sqrt{400}}{8} = \frac{\pm 20}{8}$$

Breaking up the fraction yields $x = \frac{+20}{8} = \frac{5}{2}$ or $x = \frac{-20}{8} = -\frac{5}{2}$.

As you can see, this would have been much easier to do by factoring as the difference between two squares.

5. $z = -5 + \sqrt{33}$, $z = -5 - \sqrt{33}$ The coefficients are $a = 1$, $b = 10$, and $c = -8$.

Substituting into the formula,

$$z = \frac{-10 \pm \sqrt{10^2 - 4(1)(-8)}}{2(1)} = \frac{-10 \pm \sqrt{100 - (-32)}}{2} = \frac{-10 \pm \sqrt{132}}{2} =$$
$$\frac{-10 \pm \sqrt{4}\sqrt{33}}{2} = \frac{-10 \pm 2\sqrt{33}}{2} = -5 \pm \sqrt{33}$$

Solving Quadratic Equations Using Completing the Square

As mentioned in the introduction to this section on solving quadratic equations, you'll never choose to use completing the square to solve a quadratic equation if you can factor it or use the quadratic formula. But completing the square is used when rewriting equations of conic sections (circles, ellipses, hyperbolas, and parabolas) in a form that can be easily analyzed and graphed. In this setting, using completing the square to solve a quadratic equation, you'll get the same solutions you would get using other methods, so at least you'll have an answer.

The procedure for using completing the square on the quadratic equation $ax^2 + bx + c = 0$ is

1. If a isn't 1, then divide every term in the equation by a.

2. Subtract c (or add the opposite of c) from each side of the equation.

3. Determine what new constant to add to each side of the equation so that the left side will become a perfect square trinomial. To find that constant, divide the coefficient b (or b/a) by 2 and square that result. Add it to each side.

4. Factor the left side (it's now a perfect square trinomial).

5. Take the square root of each side.

6. Solve for the variable.

Example Problems

These problems show answers and solutions.

1. Solve $x^2 + 6x - 7 = 0$ using completing the square.

   answer: $x = 1$, $x = -7$ Here are the steps:

   1. Since a is 1, you don't need to divide by anything.

   2. Add 7 to each side.

      $x^2 + 6x - 7 = 0$ becomes $x^2 + 6x = 7$

   3. Complete the square by dividing 6 by 2, squaring that, and adding it to each side.

      $x^2 + 6x + 9 = 7 + 9$

      $x^2 + 6x + 9 = 16$

   4. Factor the left side.

      $x^2 + 6x + 9 = (x + 3)(x + 3) = (x + 3)^2$

      So, $(x + 3)^2 = 16$

   5. Take the square root of each side.

      $\sqrt{(x + 3)^2} = \pm\sqrt{16}$

      Don't forget the $\pm$ signs.

      $x + 3 = \pm 4$

   6. Solve for x by subtracting 3 from each side.

      $x = -3 \pm 4$

      $x = -3 + 4 = 1$ or $x = -3 - 4 = -7$

      Yes, this would have been much easier and quicker using factoring, but it's a good example to start with when explaining completing the square.

2. Solve $3y^2 - 8y - 1 = 0$ using completing the square.

   answer: $y = \dfrac{4 + \sqrt{19}}{3}$ or $y = \dfrac{4 - \sqrt{19}}{3}$

   1. Since a is 3, divide each term by 3.

      $\dfrac{3y^2}{3} - \dfrac{8y}{3} - \dfrac{1}{3} = \dfrac{0}{3}$

      $y^2 - \dfrac{8}{3}y - \dfrac{1}{3} = 0$

   2. Add $\dfrac{1}{3}$ to each side.

      $y^2 - \dfrac{8}{3}y - \dfrac{1}{3} = 0$ becomes $y^2 - \dfrac{8}{3}y = \dfrac{1}{3}$

3. Dividing $-\frac{8}{3}$ by 2 is the same as multiplying by $\frac{1}{2}$; $-\frac{8}{3} \cdot \frac{1}{2} = -\frac{8}{6} = -\frac{4}{3}$.

   Now, $-\frac{4}{3}$ is squared and added to each side of the equation.

   $$y^2 - \frac{8}{3}y + \frac{16}{9} = \frac{1}{3} + \frac{16}{9}$$
   $$y^2 - \frac{8}{3}y + \frac{16}{9} = \frac{19}{9}$$

4. Factoring the left side of the equation,

   $$\left(y - \frac{4}{3}\right)^2 = \frac{19}{9}$$

5. Taking the square root of each side,

   $$\sqrt{\left(y - \frac{4}{3}\right)^2} = \sqrt{\frac{19}{9}}$$
   $$y - \frac{4}{3} = \pm\sqrt{\frac{19}{9}} = \pm\frac{\sqrt{19}}{3}$$

6. To solve for y, add $\frac{4}{3}$ to each side.

   $$y = \frac{4}{3} \pm \frac{\sqrt{19}}{3}, \text{ so } y = \frac{4 + \sqrt{19}}{3} \text{ or } y = \frac{4 - \sqrt{19}}{3}$$

3. Derive the quadratic formula by using completing the square on the standard form of the quadratic equation, $ax^2 + bx + c = 0$.

   result: $x = \dfrac{-b \pm \sqrt{b^2 - 4ac}}{2a}$

1. Divide every term by a.

   $$\frac{\cancel{a}x^2}{\cancel{a}} + \frac{bx}{a} + \frac{c}{a} = x^2 + \frac{b}{a}x + \frac{c}{a} = 0$$

2. Subtract $\frac{c}{a}$ from each side.

   $$x^2 + \frac{b}{a}x = -\frac{c}{a}$$

3. Add $\frac{b^2}{4a^2}$ to each side. This comes from taking half of $\frac{b}{a}$ and squaring it.

   $$x^2 + \frac{b}{a}x + \frac{b^2}{4a^2} = -\frac{c}{a} + \frac{b^2}{4a^2} = -\frac{c}{a} \cdot \frac{4a}{4a} + \frac{b^2}{4a^2} = \frac{-4ac}{4a^2} + \frac{b^2}{4a^2} = \frac{b^2 - 4ac}{4a^2}$$

4. Factor the left side.

   $$\left(x + \frac{b}{2a}\right)^2 = \frac{b^2 - 4ac}{4a^2}$$

5. Take the square root of each side.

   $$\sqrt{\left(x + \frac{b}{2a}\right)^2} = \sqrt{\frac{b^2 - 4ac}{4a^2}}$$
   $$x + \frac{b}{2a} = \pm\sqrt{\frac{b^2 - 4ac}{4a^2}} = \pm\frac{\sqrt{b^2 - 4ac}}{2a}$$

6. Solve for x by subtracting $\frac{b}{2a}$ from each side.

   $$x = -\frac{b}{2a} \pm \frac{\sqrt{b^2 - 4ac}}{2a} = \frac{-b \pm \sqrt{b^2 - 4ac}}{2a}$$

Work Problems

Use these problems to give yourself additional practice. Solve each of the quadratic equations using completing the square.

1. $y^2 + 3y - 10 = 0$

2. $2z^2 - 11z + 15 = 0$

3. $m^2 + 5m - 2 = 0$

4. $4x^2 - 49 = 0$

5. $y^2 - 6y = 0$

Worked Solutions

1. **$y = 2, y = -5$**   Add 10 to each side, $y^2 + 3y = 10$.

   Add the square of half the coefficient of y to each side, $y^2 + 3y + \left(\dfrac{3}{2}\right)^2 = 10 + \left(\dfrac{3}{2}\right)^2$.

   $$y^2 + 3y + \frac{9}{4} = 10 + \frac{9}{4} = \frac{49}{4}$$

   Factor the left side, $\left(y + \dfrac{3}{2}\right)^2 = \dfrac{49}{4}$.

   Take the square root of each side, $\sqrt{\left(y + \dfrac{3}{2}\right)^2} = \sqrt{\dfrac{49}{4}}$.

   $$y + \frac{3}{2} = \pm\sqrt{\frac{49}{4}} = \pm\frac{7}{2}$$

   Subtract $\dfrac{3}{2}$ from each side, $y = -\dfrac{3}{2} \pm \dfrac{7}{2} = \dfrac{-3 \pm 7}{2}$.

   So, $y = \dfrac{-3+7}{2} = \dfrac{4}{2} = 2$ or $y = \dfrac{-3-7}{2} = \dfrac{-10}{2} = -5$

2. **$z = 3, z = \dfrac{5}{2}$**   First divide each term by 2; then subtract $\dfrac{15}{2}$ from each side,

   $$\frac{2z^2}{2} - \frac{11z}{2} + \frac{15}{2} = z^2 - \frac{11}{2}z + \frac{15}{2} = 0.$$

   $$z^2 - \frac{11}{2}z = -\frac{15}{2}$$

   $$z^2 - \frac{11}{2}z + \left(\frac{11}{4}\right)^2 = -\frac{15}{2} + \left(\frac{11}{4}\right)^2$$

   $$z^2 - \frac{11}{2}z + \frac{121}{16} = \frac{-120}{16} + \frac{121}{16} = \frac{1}{16}$$

   $$\left(z - \frac{11}{4}\right)^2 = \frac{1}{16}$$

   $$\sqrt{\left(z - \frac{11}{4}\right)^2} = \sqrt{\frac{1}{16}}$$

$$z - \frac{11}{4} = \pm\sqrt{\frac{1}{16}} = \pm\frac{1}{4}$$

$$z = \frac{11}{4} \pm \frac{1}{4} = \frac{11 \pm 1}{4}$$

So, $z = \dfrac{11 + 1}{4} = \dfrac{12}{4} = 3$ or $z = \dfrac{11 - 1}{4} = \dfrac{10}{4} = \dfrac{5}{2}$

3. $m = \dfrac{-5 + \sqrt{33}}{2}, \ m = \dfrac{-5 - \sqrt{33}}{2}$

$$m^2 + 5m + \left(\frac{5}{2}\right)^2 = 2 + \left(\frac{5}{2}\right)^2$$

$$m^2 + 5m + \frac{25}{4} = \frac{8}{4} + \frac{25}{4} = \frac{33}{4}$$

$$\left(m + \frac{5}{2}\right)^2 = \frac{33}{4}$$

$$\sqrt{\left(m + \frac{5}{2}\right)^2} = \sqrt{\frac{33}{4}}$$

$$m + \frac{5}{2} = \pm\sqrt{\frac{33}{4}} = \pm\frac{\sqrt{33}}{2}$$

$$m = -\frac{5}{2} \pm \frac{\sqrt{33}}{2} = \frac{-5 \pm \sqrt{33}}{2}$$

4. $x = \dfrac{7}{2}, \ x = -\dfrac{7}{2}$

$$\frac{4x^2}{4} - \frac{49}{4} = x^2 - \frac{49}{4} = 0$$

$$x^2 = \frac{49}{4}$$

In this case, you don't need to add anything to each side to complete the square; the left side is already a square, so that part is done. Now, just take the square root of each side.

$$\sqrt{x^2} = \sqrt{\frac{49}{4}}$$

$$x = \pm\frac{7}{2}$$

5. **y = 6, y = 0**   There is no constant to move over to the right, so completing the square can proceed from here.

$$y^2 - 6y + (3)^2 = 0 + (3)^2$$
$$y^2 - 6y + 9 = 9$$

Factoring the left side,

$$(y - 3)^2 = 9$$

$$\sqrt{(y - 3)^2} = \sqrt{9}$$

$$y - 3 = \pm 3$$

$$y = 3 \pm 3$$

So, $y = 3 + 3 = 6$ or $y = 3 - 3 = 0$

Solving Quadratic-Like Equations

In an earlier section of this chapter, there's a description of factoring quadratic-like expressions. In this section, solving equations that contain quadratic-like expressions is covered. Solving some higher-order equations (those with exponents greater than 2) and some equations with negative or fractional exponents can be accomplished fairly easily if the terms in the equation are quadratic-like. The same techniques that are applied when solving quadratic equations can be used to help solve quadratic-like equations. There's just an extra step or two at the end, after applying the factoring to finish the particular type of equation. Quadratic-like equations are of the form $ax^{2n} + bx^n + c = 0$. To solve them, either factor using the same patterns as in factoring quadratic equations or use the quadratic formula. Then, to finish it off, solve for the specific variable.

Example Problems

These problems show the answers and solutions.

1. Solve for x in $x^4 - 4x^2 - 21 = 0$.

    answer: $x = \sqrt{7}$ or $x = -\sqrt{7}$

    Factor this into the product of two binomials using un-FOIL, $(x^2 - 7)(x^2 + 3) = 0$.

    Setting the factors equal to 0, $x^2 - 7 = 0$ or $x^2 + 3 = 0$.

    Solving $x^2 - 7 = 0$ gives $x^2 = 7$ and $x = \pm\sqrt{7}$.

    $x^2 + 3 = 0$ has no real solutions, because $\sqrt{x^2} = \sqrt{-3}$ asks for the square root of a negative number. To solve this further, see Chapter 7 on "Complex Numbers."

    The only two real solutions of this equation are $x = \sqrt{7}$ or $x = -\sqrt{7}$.

2. Solve for z in $z^{1/3} - 5z^{1/6} + 4 = 0$.

    answer: $z = 4096$ or $z = 1$

    Factoring, $\left(z^{1/6} - 4\right)\left(z^{1/6} - 1\right) = 0$.

    Now, setting the two binomials equal to 0, if $z^{1/6} - 4 = 0$ then $z^{1/6} = 4$. This time, to solve for z, raise each side to the sixth power.

    $$\left(z^{1/6}\right)^6 = \left(4\right)^6, z = 4096$$

    and if $z^{1/6} - 1 = 0$ then $z^{1/6} = 1$. Raise each side to the sixth power: $\left(z^{1/6}\right)^6 = \left(1\right)^6, z = 1$.

    The two solutions are $z = 4096$ or $z = 1$.

3. Solve for w in $w^{-2} + 2w^{-1} - 15 = 0$.

    answer: $w = -\dfrac{1}{5}$ or $w = \dfrac{1}{3}$

    This can be factored into the product of two binomials. Sometimes the exponents get a bit distracting. You can see that they fit the pattern, but it's hard to tell how to factor them.

If that happens, just rewrite the equation as a standard quadratic equation. For instance, rewrite this problem as $x^2 + 2x - 15 = 0$, substituting x's for the w^{-1}'s and x^2's for the w^{-2}'s. This new equation factors into $(x + 5)(x - 3) = 0$. Now apply this same pattern on the equation with the negative exponents.

$$w^{-2} + 2w^{-1} - 15 = (w^{-1} + 5)(w^{-1} - 3) = 0$$

If the first factor equals 0, $w^{-1} + 5 = 0$ means $w^{-1} = -5$. Write both sides as fractions, $\frac{1}{w} = \frac{-5}{1}$. Now, solving for w, $\frac{w}{1} = \frac{1}{-5}$ or $w = -\frac{1}{5}$.

If the second factor equals 0, $w^{-1} - 3 = 0$ means $w^{-1} = 3$. Write both sides as fractions, $\frac{1}{w} = \frac{1}{3}$. Now, solving for w, $\frac{w}{1} = \frac{1}{3}$ or $w = \frac{1}{3}$.

4. Solve for t in $t^4 + 2t^2 - 1 = 0$.

answer: $t = \sqrt{-1 + \sqrt{2}}$, or $t = -\sqrt{-1 + \sqrt{2}}$

This equation doesn't factor. The quadratic formula can be applied if a substitution is done, somewhat like in the preceding example.

Let y replace the t^2. Then, if $y = t^2$, $y^2 = t^4$. So the t's can be replaced.

The new equation reads $y^2 + 2y - 1 = 0$.

Now apply the quadratic formula to this equation.

$$y = \frac{-2 \pm \sqrt{4 - 4(1)(-1)}}{2(1)} = \frac{-2 \pm \sqrt{8}}{2} = \frac{-2 \pm 2\sqrt{2}}{2} = -1 \pm \sqrt{2}$$

This is what y is equal to, but you want what t is. Replace the y with t^2.

$$t^2 = -1 \pm \sqrt{2}$$

Now, take the square root of each side to get,

$$\sqrt{t^2} = \sqrt{-1 \pm \sqrt{2}}$$
$$t = \pm\sqrt{-1 \pm \sqrt{2}}$$

There are four different answers here.

$$t = +\sqrt{-1 + \sqrt{2}}, \, t = +\sqrt{-1 - \sqrt{2}}, \, t = -\sqrt{-1 + \sqrt{2}}, \text{ and } t = -\sqrt{-1 - \sqrt{2}}$$

Actually the second and fourth solutions aren't real solutions, because that's a negative value under the radical. They'll have to be eliminated.

Work Problems

Use these problems to give yourself additional practice.

1. Solve for m in $m^8 - 257m^4 + 256 = 0$.

2. Solve for z in $z^{4/3} - 13z^{2/3} = -36$.

3. Solve for p in $p^{-2} - 9 = 0$.

4. Solve for x in $3x^4 - 75x^2 = 0$.

5. Solve for y in $y^6 + 19y^3 = 216$.

Worked Solutions

1. **$m = 1$, $m = -1$, $m = 4$, $m = -4$**   Factor into the product of two binomials: $(m^4 - 1)(m^4 - 256) = 0$.

   Set the factors equal to 0. $m^4 - 1 = 0$ factors into $(m^2 - 1)(m^2 + 1) = 0$; $m^2 - 1 = 0$ factors into $(m - 1)(m + 1) = 0$. There are two real solutions from these last two factors, $m = 1$ or $m = -1$.

   $m^4 - 256 = 0$ factors into $(m^2 - 16)(m^2 + 16) = 0$; $m^2 - 16 = 0$ factors into $(m - 4)(m + 4) = 0$. There are two real solutions from these last two factors, $m = 4$ or $m = -4$.

   There are four real solutions: $m = 1$, $m = -1$, $m = 4$, $m = -4$.

2. **$z = \pm 27$, $z = \pm 8$**   Add 36 to each side. Then factor into the product of two binomials.

$$z^{4/3} - 13z^{2/3} + 36 = \left(z^{2/3} - 9\right)\left(z^{2/3} - 4\right) = 0$$

   Set the binomials equal to 0.

   When $z^{2/3} - 9 = 0$, factor as the difference of squares, $\left(z^{1/3} - 3\right)\left(z^{1/3} + 3\right) = 0$.

   When $z^{1/3} - 3 = 0$, $z^{1/3} = 3$. Cube each side, and $z = 27$.

   When $z^{1/3} + 3 = 0$, $z^{1/3} = -3$. Cube each side to get $z = -27$.

   When $z^{2/3} - 4 = 0$, factor as the difference of squares, $\left(z^{1/3} - 2\right)\left(z^{1/3} + 2\right) = 0$.

   When $z^{1/3} - 2 = 0$, $z^{1/3} = 2$. Cube each side, and $z = 8$.

   When $z^{1/3} + 2 = 0$, $z^{1/3} = -2$. Cube each side to get $z = -8$.

   The four solutions are $z = \pm 27$ or $z = \pm 8$.

3. $p = \frac{1}{3}$, $p = -\frac{1}{3}$ Factoring the binomial using the difference of two squares,
$p^{-2} - 9 = \left(p^{-1} - 3\right)\left(p^{-1} + 3\right) = 0$.

Now set each factor equal to 0. When $p^{-1} - 3 = 0$, $p^{-1} = 3$. Writing both sides as fractions and flipping them, $\frac{1}{p} = \frac{3}{1}$, gives you $\frac{p}{1} = \frac{1}{3}$ or $p = \frac{1}{3}$.

When $p^{-1} + 3 = 0$, $p^{-1} = -3$. Writing both sides as fractions and flipping them, $\frac{1}{p} = \frac{-3}{1}$, gives you $\frac{p}{1} = \frac{1}{-3}$ or $p = -\frac{1}{3}$.

The solutions are then $p = \pm\frac{1}{3}$.

4. $x = 0$, $x = 5$, $x = -5$ First factor the binomial by dividing by the Greatest Common Factor, $3x^2$.

$$3x^4 - 75x^2 = 3x^2(x^2 - 25) = 0$$

The binomial can be factored, and then each of the factors is set equal to 0.
$3x^2(x - 5)(x + 5) = 0$.

When $x^2 = 0$, $x = 0$; when $x - 5 = 0$, $x = 5$; when $x + 5 = 0$, $x = -5$.

The solutions are then $x = 0, \pm 5$.

5. $y = -3$, $y = 2$ Subtract 216 from each side. Then factor the quadratic-like trinomial.

$y^6 + 19y^3 - 216 = 0$ becomes $(y^3 + 27)(y^3 - 8) = 0$. Setting each factor equal to 0, when $y^3 + 27 = 0$, $y^3 = -27$ and $y = -3$. When $y^3 - 8 = 0$, $y^3 = 8$ and $y = 2$.

Quadratic and Other Inequalities

A quadratic inequality is an inequality with a quadratic or second-degree term in it and no higher powers. Solving quadratic inequalities takes a bit of special handling. As soon as a second-degree or higher-degree term gets involved, the Multiplication Property of Zero comes into play. The same is true when fractions are in an inequality. The general plan for dealing with these problems is to get all the terms on one side of the inequality symbol and completely factor the expression. The Multiplication Property of Zero determines the values where the factors might change signs (negative to positive or positive to negative). A solution is determined by whether the expression should be positive or negative, so it's important to know when which factor will be which sign and when. For example, the inequality $x^2 - 4x > 0$ is true when the left side is positive. If $x = 10$, then the left side of the inequality is $100 - 40$, which is positive. That means that the number 10 is part of the solution of the inequality. If $x = 1$, then the left side of the inequality is $1 - 4$, which is -3. This is negative, not positive, so the number 1 is not a part of the solution. If $x = -10$, then the left side of the inequality is $100 + 40$, which is positive, so -10 is part of the solution. These numbers that work seem to be all over the place! But really, any negative number is part of the solution, and any number bigger than 4 works, too. The examples that follow will show you how to solve the inequalities.

Example Problems

These problems show the answers and solutions.

1. Solve $y^2 + 7y - 8 > 0$.

   answer: $y < -8$ or $y > 1$

   The solution consists of all values of y that make the left side positive.

   Factor the left side into $(y + 8)(y - 1) > 0$. Now find out where the factors change sign by determining where they equal 0. $y + 8$ changes sign at -8 because numbers smaller than -8 make the expression negative, and numbers larger than -8 make the expression positive.

   $y - 1$ changes sign at 1. Now graph these two values on a number line.

   Put $+$ and $-$ in the parentheses above the number line to show whether the factors are $+$ or $-$ in that interval.

   For numbers less than -8, for instance, choose $y = -10$, the factor $y + 8$ is negative, and the factor $y - 1$ is negative. Put this on the number line to represent $(y + 8)(y - 1)$,

   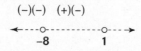

   Now check on those two factors for numbers between -8 and 1. A good number to choose is $y = 0$. The factor $y + 8$ is positive, and the factor $y - 1$ is negative. Put this on the number line to represent $(y + 8)(y - 1)$,

   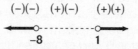

   The last place to check is for values bigger than 1. Letting $y = 2$, both factors are positive, so

   $$(-)(-) \quad (+)(-) \quad (+)(+)$$

   Going back to the original problem, which is to determine when the expression on the left is positive, just look at the factors. Two negatives multiplied together give a positive product, and two positives give a positive product. A positive times a negative is negative, so you just want the numbers smaller than -8 or bigger than 1 in your solution. This is written $y < -8$ or $y > 1$.

2. Solve $x^2(x + 3)(x - 8) \leq 0$.

    answer: $-3 \leq x \leq 8$

    There are three factors to consider. Their product has to be either negative or equal to 0. The Multiplication Property of Zero takes care of the zeros. The factors are equal to 0 at $x = 0$, $x = -3$, and $x = 8$. Now all that has to be done is determine the signs of the factors between the numbers that make the factors 0. The number line will contain the three values already determined.

    Choose a number smaller than -3, for example, $x = -4$. Putting that into the factors $x^2(x + 3)(x - 8)$ yields $(+)(-)(-)$. That will go to the left of the -3.

    Choose a number between -3 and 0, for example, $x = -1$. Putting that into the factors $x^2(x + 3)(x - 8)$ yields $(+)(+)(-)$. That will go between the -3 and 0.

    Choose a number between 0 and 8, for example, $x = 1$. Putting that into the factors $x^2(x + 3)(x - 8)$ yields $(+)(+)(-)$. That will go between the 0 and 8.

    Choose a number bigger than 8, for example, $x = 10$. Putting that into the factors $x^2(x + 3)(x - 8)$ yields $(+)(+)(+)$. That will go to the right of the 8.

    $(+)(-)(-)$ $(+)(+)(-)$ $(+)(+)(-)$ $(+)(+)(+)$

    Now to go back to what the original problem was. Where are the x's that make the expression 0 or negative? Whenever an odd number of negative factors exists, the whole product is negative, so the numbers between -3 and 8 will be in the solution, including the -3 and the 8, because they make the expression equal to 0. The answer is written $-3 \leq x \leq 8$.

3. Solve for z in $\dfrac{z + 5}{3 - z} \geq 0$.

    answer: $-5 \leq z < 3$

    Even though this is a fraction, the numerator and denominator can be treated the same way as the factors in a multiplication problem. Determine where they change sign and use that to find out when the expression is positive or 0. The numerator is 0 and changes sign when $z = -5$. The denominator is 0 and changes sign when $z = 3$. So, make a number line with those two numbers on it.

    This time stack the signs. When you choose a number smaller than -5, such as $z = -6$, the factor $z + 5$ is negative and the factor $3 - z$ is positive. Put $\dfrac{-}{+}$ on the number line.

When you choose a number between −5 and 3, such as $z = 0$, the factor $z + 5$ is positive and the factor $3 − z$ is positive. Put $\frac{+}{+}$ on the number line.

When you choose a number bigger than 3, such as $z = 4$, the factor $z + 5$ is positive and the factor $3 − z$ is negative. Put $\frac{+}{-}$ on the number line.

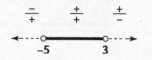

You're looking for where the result is positive or 0. The numbers between −5 and 3 will always give a positive result, so they're part of the solution. The number −5 makes the numerator equal to 0, so the whole fraction is equal to 0. It can be part of the solution. You can't use the 3, however, even though it makes a factor 0. The fact that it makes the denominator 0 is not allowed. No number can be the result of dividing by 0.

So the solution is written $−5 \le z < 3$.

Work Problems

Use these problems to give yourself additional practice.

1. Solve for x in $x^2 + 6x − 27 > 0$.

2. Solve for z in $4z^2 + 4z − 15 \le 0$.

3. Solve for m in $m(m − 3)(m + 5)(m − 7) \ge 0$.

4. Solve for n in $\dfrac{1}{3 + n} > 0$.

5. Solve for p in $\dfrac{p − 8}{p − 5} \le 0$.

Worked Solutions

1. **$x < −9$ or $x > 3$**   Factor the trinomial, $(x + 9)(x − 3) > 0$.

    The values where the factors change sign are $x = −9$ and $x = 3$.

    Make a number line and indicate the signs of the factors in those intervals.

    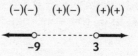

    The values that make the left side positive are $x < −9$ or $x > 3$.

2. $-\dfrac{5}{2} \leq z \leq \dfrac{3}{2}$ Factor the trinomial, $(2z - 3)(2z + 5) \leq 0$.

The values where the factors change sign are $z = \dfrac{3}{2}$ and $z = -\dfrac{5}{2}$.

Make a number line and indicate the signs of the factors in those intervals.

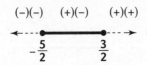

The values which make the left side negative or 0 are $-\dfrac{5}{2} \leq z \leq \dfrac{3}{2}$.

3. $\mathbf{0 \leq m \leq 3,\ m \geq 7,\ or\ m \leq -5}$ The values where the factors change sign are $m = 0, 3,$ -5, and 7.

Make a number line and indicate the signs of the factors in those intervals.

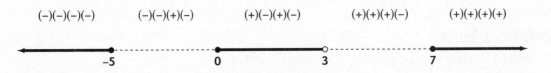

The values which make the left side positive or 0 are $m \leq -5$ or $0 \leq m \leq 3$ or $m \geq 7$.

4. $\mathbf{n > -3}$ The only value for which a factor changes sign is -3, for the factor in the denominator.

If n is a number bigger than -3, then the denominator is positive. If n is smaller than -3, the denominator is negative. The denominator cannot be 0, so n is never equal to -3. The solutions, which make the fraction positive, are the numbers larger than -3. $n > -3$.

5. $\mathbf{5 < p \leq 8}$ The values where the factors change sign are $p = 8$ and $p = 5$.

Make a number line and indicate the signs of the factors in those intervals.

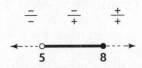

The values that make the left side negative or 0 are $5 < p \leq 8$. You can use the 8, because that makes the numerator equal to 0. The denominator is 0 when p is 5, so that value can't be part of the solution.

Radical Equations with Quadratics

Equations including radical expressions are most easily solved by first getting rid of the radical. When changing from radical equations to *linear* equations, the new equation is easily solved in that format. In this section, you'll see how to handle the special problems that can arise when you change from a radical form to a *quadratic* equation. There are opportunities for false or extraneous solutions to arise. The solution of the corresponding quadratic equation can result in two solutions to the radical equation, just one solution, or none at all. The important thing to do when solving those radical equations is to be sure to check your final answer. The procedure will be to get the radical term on one side of the equation and the rest of the terms on the other side. Then square both sides and solve the resulting equation. And, the key to getting a correct solution is to check your answers.

Example Problems

These problems show the answers and solutions.

1. Solve $\sqrt{2x + 20} = x - 2$.

    answer: $x = 8$

    Square both sides, $\left(\sqrt{2x + 20}\right)^2 = (x - 2)^2$.

    The left side will be the binomial under the radical. The right side will be the perfect square binomial that you get from multiplying the two binomials together.

    $$2x + 20 = x^2 - 4x + 4$$

    Set the quadratic equal to zero by subtracting $2x$ and 20 from each side.

    $$0 = x^2 - 6x - 16$$

    Factor and solve for x: $0 = (x - 8)(x + 2)$, so $x = 8$ or $x = -2$.

    Now these both need to be checked. They're solutions of the quadratic equation but not necessarily of the radical equation.

    When $x = 8$, $\sqrt{2(8) + 20} = (8) - 2$; $\sqrt{36} = 6$. This is true. The number 8 is a solution.

    When $x = -2$, $\sqrt{2(-2) + 20} = (-2) - 2$; $\sqrt{16} \neq -4$. This one does not work. So the only solution is $x = 8$.

2. Sometimes, both solutions of the quadratic work. Solve $\sqrt{5x - 1} = x + 1$.

    answer: $x = 2$, $x = 1$

    Squaring both sides and setting the quadratic equal to 0,

    $$\left(\sqrt{5x - 1}\right)^2 = (x + 1)^2$$
    $$5x - 1 = x^2 + 2x + 1$$
    $$0 = x^2 - 3x + 2$$

    Now factor and solve for the two solutions of the quadratic.

    $$0 = (x - 2)(x - 1)$$
    $$x = 2 \text{ or } x = 1$$

Checking these in the original equation:

If $x = 2$, $\sqrt{5(2) - 1} = (2) + 1$, $\sqrt{9} = 3$. This one works.

If $x = 1$, $\sqrt{5(1) - 1} = (1) + 1$, $\sqrt{4} = 2$. This one also works.

Work Problems

Use these problems to give yourself additional practice.

1. Solve $\sqrt{x + 20} = x$.

2. Solve $\sqrt{z^2 + 9} = 2z - 3$.

3. Solve $\sqrt{3n + 1} + 3 = n$.

4. Solve $\sqrt{t^2 + 25} = t + 5$.

5. Solve $\sqrt{8x + 1} = x + 2$.

Worked Solutions

1. **$x = 5$**   Square both sides and set the quadratic equal to 0.

$$\left(\sqrt{x + 20}\right)^2 = x^2$$
$$x + 20 = x^2$$
$$0 = x^2 - x - 20$$

Factoring and solving for x, $0 = (x - 5)(x + 4)$, which means that $x = 5$ or $x = -4$.

If $x = 5$, then $\sqrt{5 + 20} = 5$ or $\sqrt{25} = 5$. The 5 works.

If $x = -4$, then $\sqrt{-4 + 20} = -4$ or $\sqrt{16} \neq -4$. The -4 doesn't work.

2. **$z = 4$**   Square both sides and set the quadratic equal to 0.

$$\left(\sqrt{z^2 + 9}\right)^2 = (2z - 3)^2$$
$$z^2 + 9 = 4z^2 - 12z + 9$$
$$0 = 3z^2 - 12z$$
$$0 = 3z(z - 4)$$

The values that make this true are $z = 0$ or $z = 4$.

Testing $z = 0$, $\sqrt{(0)^2 + 9} = 2(0) - 3$ or $3 = -3$. The 0 doesn't work.

Testing $z = 4$, $\sqrt{(4)^2 + 9} = 2(4) - 3$ or $\sqrt{25} = 5$. The 4 works.

3. **$n = 8$** First subtract 3 from each side. Then square both sides.

$$\sqrt{3n + 1} = n - 3$$
$$\left(\sqrt{3n + 1}\right)^2 = (n - 3)^2$$
$$3n + 1 = n^2 - 6n + 9$$

Set the quadratic equal to zero, factor it, and determine the solutions.

$0 = n^2 - 9n + 8$, $0 = (n - 1)(n - 8)$. The solutions are $n = 1$ or $n = 8$.

Testing $n = 1$, $\sqrt{3(1) + 1} + 3 = 1$ or $\sqrt{4} + 3 = 1$ or $2 + 3 = 1$. The 1 does not work

Testing $n = 8$, $\sqrt{3(8) + 1} + 3 = 8$ or $\sqrt{25} + 3 = 8$ or $5 + 3 = 8$. The 8 does work.

The solution of the radical equation is just $n = 8$.

4. **$t = 0$** Square both sides and solve the quadratic equation.

$$\left(\sqrt{t^2 + 25}\right)^2 = (t + 5)^2$$
$$t^2 + 25 = t^2 + 10t + 25$$

Subtract t^2 and 25 from each side.

$$0 = 10t \text{ or } t = 0$$

Testing this in the original equation, $\sqrt{0 + 25} = 0 + 5$ or $\sqrt{25} = 5$. The only solution is $t = 0$.

5. **$x = 1, x = 3$** Square both sides and solve the quadratic equation.

$$\left(\sqrt{8x + 1}\right)^2 = (x + 2)^2$$
$$8x + 1 = x^2 + 4x + 4$$
$$0 = x^2 - 4x + 3$$
$$0 = (x - 1)(x - 3) \text{ so } x = 1 \text{ or } x = 3.$$

Testing $x = 1$, $\sqrt{8(1) + 1} = (1) + 2$ or $\sqrt{9} = 1 + 2$. The 1 works.

Testing $x = 3$, $\sqrt{8(3) + 1} = (3) + 2$ or $\sqrt{25} = 3 + 2$. The 3 works, also. There are two solutions for this radical equation, $x = 1$ or $x = 3$.

Chapter 3
Function Operations and Transformations

Functions are very specific mathematical relationships. They can be manipulated and operated upon using most of our standard operations of addition, subtraction, and so on. They also have some operations that are very special to themselves, only. The results of performing these operations upon functions are new functions—some with the same properties as the original and some that have totally different characteristics. All of these processes are discussed in this chapter.

Functions and Function Notation

You can express the relationship between two variables in many ways. The usual ways of doing this are pictures (graphs), tables of numbers, and formulas with equations or inequalities. A particular type of relationship that is considered most frequently in algebra is a relationship called a *function*. The thing that sets a function aside from all other relationships between two variables is that one of the variables is said to be a function of the other if the operations involved in the equation always produce a *unique* result.

The equation $y = x^2 + 3$ represents a function where "y is a function of x." The equation $y^2 = x^2 + 16$ is **not** a function, even though it represents a relationship between the variables y and x. What distinguishes a function from just any old relation? When y is a function of x, it's "driven" by x. The x variables are the input variables, and the y variables are the output variables. When y is a *function* of x, there is *exactly one* output value for every input value. There's exactly one y for every x.

A main concern of whether or not an equation represents a function involves its usefulness as a formula or model in a practical situation. If an equation or formula is supposed to tell a store owner how much it costs to stock x number of an item, then she wants the formula to have one answer for any x that's input. It isn't very helpful if the formula gives two different answers.

Domain and Range

When studying functions, for instance those in which y is a function of x, some properties and characteristics are important when choosing and using them. The *domain* and *range* of a function are such properties.

Domain

The domain of a function contains all of the possible *input* values that you can use—every number that can be put into the formula or equation and get a real answer. The function $y = \sqrt{x-2}$ has a domain that contains the number 2 and every number greater than 2, written $x \geq 2$. The domain can't include anything smaller than 2, because then there would be a negative number under the radical, and that kind of number isn't real.

Range

The range of a function contains all of the possible *output* values—every number that is a result of putting input values into the formula or equation. The function $y = \sqrt{x-2}$ has a range that contains the number 0 and also contains every positive number. The range doesn't include any negatives, because this radical operation results in only positive answers or 0.

When determining the domain and range of functions, just a few operations cause restrictions or special attention. Functions with radicals that have even roots will have restricted domains. You can't take the square root or fourth root of a negative number, so any number that would create that situation has to be eliminated from the domain. Fractions also have to be considered carefully. Any x value that creates a 0 in the denominator has to be eliminated from the domain. Functions with even powered radicals or absolute value will have restricted ranges. They will produce just positive results unless they're multiplied by -1; then everything would be changed to the opposite sign. Some other special cases exist, but they'll have to be determined by trying a few coordinates or by putting x values into the function equation.

Function Notation

First, look at some notation for functions. The notation used so far has been in terms of x's and y's, and, usually, y is a function of x. The functions look like $y = x + 8$ or $y = \sqrt{4x-1}$, and so on. A common notation, called *function notation*, writes these last two functions as $f(x) = x + 8$ or $g(x) = \sqrt{4x-1}$. This is read, "f of x equals x plus 8" or "g of x equals the square root of $4x$ minus 1." This function notation allows you to name functions with the letters f, g, h, and so on. It also is handy when writing down coordinates of points or numbers from a chart. If you're considering the function $f(x) = x + 8$ and want to know what the output is if $x = 2$, write $f(2) = 2 + 8 = 10$. The input is 2; the output is 10.

Example Problems

These problems show the answers and solutions.

1. Determine the domain and range of the function $f(x) = x^2$.

    answer: Domain: all real numbers. Range: all positive numbers or 0, $f(x) \geq 0$.

    The domain, all the values that x can be, are all real numbers. No matter what number you put in for x, you'll get a real answer. The range, or the results of putting numbers in for x, consists of only positive numbers or 0. When you square either a positive or negative number, the result will always be positive.

2. Determine the domain and range of the function $g(x) = \frac{1}{x}$.

   answer: Domain: all numbers except 0. Range: all numbers except 0.

   The x values can be anything except 0. The number 0 can't divide into anything, because there's no answer to that problem. Also, the range won't include 0 either, because, for a fraction to equal 0, the numerator has to be 0. The numerator is always the constant 1, so that just won't happen.

3. Determine the domain and range of the function $h(x) = \sqrt{x - 2}$.

   answer: Domain: all real numbers bigger than or equal to 2, $x \geq 2$. Range: all positive numbers or 0, $h(x) \geq 0$.

   The radical is a square root—the root is an even number, so a negative number can't be under the radical sign. This won't happen unless the x value is smaller than 2, so choosing 2 or something bigger works. The range or results will always be positive as long as x is bigger than 2. The result will be 0 when x equals 2.

4. If the function $k(x) = 2x^2 - 3x + 1$, find the function values for the following: $k(2)$, $k(-3)$, $k(0)$.

   answer: $k(2) = 3$; $k(-3) = 28$; $k(0) = 1$

   $k(2) = 2(2)^2 - 3(2) + 1 = 8 - 6 + 1 = 3$. Each x is replaced with a 2, and the operation is performed on the 2.

   $k(-3) = 2(-3)^2 - 3(-3) + 1 = 18 + 9 + 1 = 28$. Each x is replaced with a -3, and the operation is performed on it.

   $k(0) = 2(0)^2 - 3(0) + 1 = 1$. Replacing the x's with 0's left only the constant, 1.

Work Problems

Use these problems to give yourself additional practice.

1. Find the domain and range of the function $f(x) = 8 - x^2$.

2. Find the domain and range of the function $h(x) = \sqrt{x - 3}$.

3. Find the domain and range of the function $m(x) = \frac{3}{x - 1}$.

4. Given the function $h(x) = \sqrt{x - 3}$, find the function values $h(4)$, $h(9)$, $h(28)$.

5. Given the function $m(x) = \frac{3}{x - 1}$, find the function values $m(2)$, $m(-5)$, $m(1)$.

Worked Solutions

1. **Domain: all real numbers; range: all numbers 8 or smaller, $f(x) \leq 8$.** The domain is all real numbers because any number can be squared and then subtracted. The range has to be 8 or less because a positive number or 0 is always subtracted from 8. The very largest the result can be is 8, which occurs when $x = 0$.

2.   **Domain: all numbers 3 or larger, $x \geq 3$; range: all positive numbers or 0, $h(x) \geq 0$.**
     The number under the radical has to be positive or 0, so no numbers smaller than 3 can be used in the domain. The results of taking the square root of a positive number or 0 is a positive number or 0.

3.   **Domain: any number except 1; range: any number except 0.** The domain can't contain the number 1, because that would result in a 0 in the denominator—there's no such number. The range will never include 0, because there can't be a 0 in the numerator. That's the only way a fraction can be equal to 0—to have the 0 in the numerator.

4. $h(4) = 1$, $h(9) = \sqrt{6}$, $h(28) = 5$ Given the function $h(x) = \sqrt{x-3}$,
     $h(4) = \sqrt{4-3} = \sqrt{1} = 1$; $h(9) = \sqrt{9-3} = \sqrt{6}$; $h(28) = \sqrt{28-3} = \sqrt{25} = 5$

5. $m(2) = 3$, $m(-5) = -\frac{1}{2}$, $m(1)$ **doesn't exist.**   Given the function $m(x) = \dfrac{3}{x-1}$,
     $m(2) = \dfrac{3}{2-1} = \dfrac{3}{1} = 3$; $m(-5) = \dfrac{3}{-5-1} = \dfrac{3}{-6} = -\dfrac{1}{2}$; $m(1) = \dfrac{3}{1-1} = \dfrac{3}{0}$
     which cannot be.

Function Operations

Functions can be added, subtracted, multiplied, and divided. The operations used to define the functions themselves are incorporated into the process used to combine the functions. The result of adding, subtracting, multiplying, or dividing functions is another, usually new, function. The domain of this new function will depend on the operations used and what the domains of the original functions were.

There's also an operation that can be used on functions, which is special to functions. It's called the *composition* operation. Its symbol is a small circle, ∘. To compose the functions $f(x)$ and $g(x)$, write $(f \circ g)(x)$. This operation isn't used on numbers. We don't have any rules for $7 \circ 6$. This operation is used to combine functions, only.

Addition, Subtraction, Multiplication, and Division

When two functions are added or subtracted, the like terms are combined using the usual algebra rules. A new function is formed containing the results of the operations. The domain of the resulting function will be what is shared by the domains of the original functions. Multiplying and dividing functions also uses the algebra rules. The domains of functions formed by dividing will incorporate the domains of the original functions and also involve restrictions based on when the function in the denominator is equal to 0.

Example Problems

These problems show the answers and solutions.

1.   Add the functions $f(x) = x^2 + 2x - 3$ and $g(x) = x^3 + 3x^2 - 2x + 4$.

     answer: $(f + g)(x) = x^3 + 4x^2 + 1$

     The like terms are combined to produce this answer. The functions f and g are polynomials whose domains are all real numbers. The same is true of the new function; its domain is all real numbers.

2. Subtract the function $h(x) = \sqrt{x} - 4$ and $k(x) = \frac{3}{x} + 2$.

answer: $(h-k)(x) = \sqrt{x} - \frac{3}{x} - 6$

When these two functions are subtracted, the only terms that combine are the two numbers: $h(x) - k(x) = (h-k)(x) = (\sqrt{x} - 4) - (\frac{3}{x} + 2) = \sqrt{x} - 4 - \frac{3}{x} - 2 = \sqrt{x} - \frac{3}{x} - 6$.

The domain is $x > 0$, or all numbers greater than 0. This new domain is a combination of the domains of the original two functions. The domain of h is all numbers 0 and greater. The domain of k excludes 0. So, their combination is what they both share—all positive numbers.

3. Divide the functions $n(x) = x^2 + 1$ and $d(x) = x^2 - 4$.

answer: $\left(\frac{n}{d}\right)(x) = \frac{x^2+1}{x^2-4}$

The denominator doesn't divide evenly into the numerator, so this is left as a fraction. No common factors exist, so the fraction can't be reduced. The denominator is equal to 0 when x is equal to 2 or −2, so those two values cannot be included in the domain of the new function. Any other real numbers work.

Work Problems

Do these problems to give yourself additional practice.

1. Add $f(x) = x^3 - 2x^2 + 7$ and $g(x) = 4x^2 + x - 2$.

2. Subtract $h(x) = \frac{x}{x-2}$ and $k(x) = \frac{2}{x-1}$.

3. Multiply $p(x) = \sqrt{x^2+1}$ and $t(x) = \sqrt{x}$.

4. Divide $m(x) = x$ and $n(x) = x - 3$.

5. Add $a(x) = x^2 + 2$ and $b(x) = x^2 + 3$ and $c(x) = x^2 + 4$.

Worked Solutions

1. **$(f+g)(x) = x^3 + 2x^2 + x + 5$** $(f)(x) + g(x) = (f+g)(x) = x^3 - 2x^2 + 7 + 4x^2 + x - 2$
The like terms are combined; the domain is all real numbers.

2. $(\mathbf{h-k})(\mathbf{x}) = \frac{x^2 - 3x + 4}{(x-2)(x-1)}$ A common denominator is first found before the terms are

combined: $h(x) = \frac{x}{x-2} \cdot \frac{x-1}{x-1} = \frac{x^2-x}{(x-2)(x-1)}$ and $k(x) = \frac{2}{x-1} \cdot \frac{x-2}{x-2} = \frac{2x-4}{(x-1)(x-2)}$.

The numerators of these new fractions are distributed so they can be combined when the two fractions are subtracted. The denominators are left in their factored form:

$$h(x) - k(x) = \frac{x^2-x}{(x-2)(x-1)} - \frac{2x-4}{(x-1)(x-2)} = \frac{x^2-x-2x+4}{(x-2)(x-1)} = \frac{x^2-3x+4}{(x-2)(x-1)}$$

The domain of the function h excluded the number 2, and the domain of the function k excluded the number 1. The domain of this new function excludes both 2 and 1. All other real numbers will work.

3. $\left(p \cdot t\right)(x) = \sqrt{x^3 + x}$ The values under the radicals can be multiplied. The domain of the function p is all real numbers, and the domain of t is all numbers 0 or greater. The resulting function also has a domain of all numbers 0 or greater.

4. $\left(\dfrac{m}{n}\right)(x) = \dfrac{x}{x-3}$ The fraction cannot be reduced. The domain excludes the number 3.

5. $(a + b + c)(x) = 3x^2 + 9$ The like terms are combined. The domain is all real numbers.

Composition of Functions

The composition of functions, the operation special to functions, is denoted with the small ∘. To perform this operation on two functions, f and g, write $(f \circ g)(x) = f(g(x))$. The way this works is that the operation designates that the second function is the input into the first function. The operations defining the first function are performed on those input values. The composition of functions is not commutative like addition and multiplication are, so the order—or which is the input—matters.

Example Problems

These problems show the answers and solutions.

1. Find $f(x) \circ g(x)$ if $f(x) = x^2 + 2x - 1$ and $g(x) = 3x - 4$.

    answer: $(f \circ g)(x) = 9x^2 - 18x + 7$

    The composition of $f \circ g = (f \circ g)(x) = f(g(x))$ means to replace every x in the rule for f with the function $g(x)$: $f(g(x)) = [g(x)]^2 + 2[g(x)] - 1 = (3x + 4)^2 + 2(3x - 4) - 1$. The operations are performed, and the result is simplified:
    $(3x - 4)^2 + 2(3x - 4) - 1 = 9x^2 - 24x + 16 + 6x - 8 - 1 = 9x^2 - 18x + 7$.

2. Find $g(x) \circ f(x)$ if $f(x) = x^2 + 2x - 1$ and $g(x) = 3x - 4$.

    answer: $(g \circ f)(x) = 3x^2 + 6x - 7$

    The composition of $g \circ f = (g \circ f)(x) = g(f(x))$ means to replace every x in the rule for g with the function $f(x)$: $(g \circ f)(x) = g(f(x)) = 3[f(x)] - 4 = 3(x^2 + 2x - 1) - 4 = 3x^2 + 6x - 3 - 4 = 3x^2 + 6x - 7$. Examples 1 and 2 illustrate that the composition of functions is not commutative.

Work Problems

Use these problems to give yourself additional practice.

1. Find $f \circ g(x)$ if $f(x) = 5x^2 - 2x + 3$ and $g(x) = 9 - 3x$.

2. Find $f \circ g(x)$ if $f(x) = \sqrt{x^2 + 4}$ and $g(x) = x^2 - 1$.

3. Find $g \circ f(x)$ if $f(x) = 5x^2 - 2x + 3$ and $g(x) = 9 - 3x$.

4. Find $f \circ g(x)$ if $f(x) = \dfrac{1}{2x - 3}$ and $g(x) = \dfrac{3x + 1}{2x}$.

5. Find $f \circ g(x)$ if $f(x) = x^2$ and $g(x) = \dfrac{1}{x}$.

Worked Solutions

1. $\mathbf{390 - 264x + 45x^2}$ The composition of the functions has the operations of the function f performed on the function g: $f \circ g(x) = f(g(x)) = 5[g(x)]^2 - 2[g(x)] + 3 = 5(9 - 3x)^2 - 2(9 - 3x) + 3 = 5(81 - 54x + 9x^2) - 18 + 6x + 3 = 405 - 270x + 45x^2 - 15 + 6x = 390 - 264x + 45x^2$.

2. $\mathbf{\sqrt{x^4 - 2x^2 + 5}}$ The composition $f \circ g(x) = f(g(x)) = \sqrt{[g(x)]^2 + 4} = \sqrt{(x^2 - 1)^2 + 4} = \sqrt{x^4 - 2x^2 + 1 + 4} = \sqrt{x^4 - 2x^2 + 5}$.

3. $\mathbf{-15x^2 + 6x}$ The composition $g \circ f(x) = g(f(x)) = 9 - 3(f(x)) = 9 - 3(5x^2 - 2x + 3) = 9 - 15x^2 + 6x - 9$.

4. $\mathbf{x}$ The composition $f \circ g(x) = f(g(x)) = \dfrac{1}{2[g(x)] - 3} = \dfrac{1}{2\left[\dfrac{3x+1}{2x}\right] - 3} = \dfrac{1}{\dfrac{6x+2}{2x} - 3}$

   $= \dfrac{1}{\dfrac{6x + 2 - 6x}{2x}} = \dfrac{2x}{6x + 2 - 6x} = \dfrac{2x}{2} = x$. In this problem, a complex fraction was created.

   The numerator was multiplied by the reciprocal of the denominator to simplify it.

5. $\mathbf{\dfrac{1}{x^2}}$ The composition $f \circ g(x) = f(g(x)) = [g(x)]^2 = \left[\dfrac{1}{x}\right]^2 = \dfrac{1}{x^2}$.

Difference Quotient

One important use of the composition of functions is in simplifying the *difference quotient* in calculus. The difference quotient is the basis of a major process in calculus, so it's important to be able to complete the algebraic part successfully. The difference quotient looks like this: $\dfrac{f(x + h) - f(x)}{h}$. This says that, if you have a function $f(x)$, then first do the function evaluation $f(x + h)$ where the input is $x + h$, subtract the function $f(x)$ from the result, simplify the numerator, and then divide the result by h.

Example Problems

These problems show the answers and solutions.

1. Find $\dfrac{f(x + h) - f(x)}{h}$ if $f(x) = 3x^2 - 2x + 9$.

   answer: $6x + 3h - 2$

   The first step is to determine $f(x + h)$. This is the same as saying that there's a composition of functions (see the previous section for more on this) $f \circ g(x)$ where $f(x) = 3x^2 - 2x + 9$ and $g(x) = x + h$.

   $f \circ g(x) = f(x + h) = 3(x + h)^2 - 2(x + h) + 9 = 3(x^2 + 2xh + h^2) - 2x - 2h + 9 = 3x^2 + 6xh + 3h^2 - 2x - 2h + 9$. This value is substituted into the difference formula.

$$\frac{f(x+h)-f(x)}{h} = \frac{3x^2 + 6xh + 3h^2 - 2x - 2h + 9 - (3x^2 - 2x + 9)}{h}$$

$$= \frac{3x^2 + 6xh + 3h^2 - 2x - 2h + 9 - 3x^2 + 2x - 9}{h}$$

$$= \frac{3\cancel{x^2} + 6xh + 3h^2 - \cancel{2x} - 2h + \cancel{9} - 3\cancel{x^2} + \cancel{2x} - \cancel{9}}{h}$$

All the terms that have been crossed out have their inverse in the numerator, too, so their sum is 0. That leaves $\frac{6xh + 3h^2 - 2h}{h} = \frac{h(6x + 3h - 2)}{h} = \frac{\cancel{h}(6x + 3h - 2)}{\cancel{h}} = 6x + 3h - 2$.
In calculus, this result is used to find the *derivative* of the function.

2. Find $\frac{f(x+h)-f(x)}{h}$ if $f(x) = \sqrt{2x - 1}$.

answer: $\dfrac{2}{\sqrt{2x + 2h - 1} + \sqrt{2x - 1}}$

First find $f(x+h) = \sqrt{2(x+h) - 1}$. Substitute this into the difference quotient,
$\frac{f(x+h)-f(x)}{h} = \frac{\sqrt{2(x+h) - 1} - \sqrt{2x - 1}}{h}$. Even though the answer may not look
any nicer than this expression, the goal is to get rid of the h in the denominator.
This can be accomplished by *rationalizing* the numerator. For a review of rationalizing
fractions, see Chapter 1.

$$\frac{\sqrt{2(x+h) - 1} - \sqrt{2x - 1}}{h} \cdot \frac{\sqrt{2(x+h) - 1} + \sqrt{2x - 1}}{\sqrt{2(x+h) - 1} + \sqrt{2x - 1}} = \frac{2x + 2h - 1 - (2x - 1)}{h\left(\sqrt{2(x+h) - 1} + \sqrt{2x - 1}\right)}$$

$$= \frac{2x + 2h - 1 - (2x - 1)}{h\left(\sqrt{2(x+h) - 1} + \sqrt{2x - 1}\right)} = \frac{2\cancel{h}}{\cancel{h}\left(\sqrt{2(x+h) - 1} + \sqrt{2x - 1}\right)} = \frac{2}{\left(\sqrt{2(x+h) - 1} + \sqrt{2x - 1}\right)}$$

Believe it or not, this is a very nice result.

Work Problems

Use these problems to give yourself additional practice. Find the difference quotient for each.

1. $f(x) = 2x - 1$

2. $f(x) = 3x^2 + x - 7$

3. $f(x) = \sqrt{x + 3}$

4. $f(x) = \dfrac{1}{x + 7}$

5. $f(x) = x^3$

Worked Solutions

1. **2** $\dfrac{f(x+h)-f(x)}{h}=\dfrac{2(x+h)-1-(2x-1)}{h}=\dfrac{2x+2h-1-2x+1}{h}=\dfrac{2h}{h}=2$

2. **6x + 3h + 1** $\dfrac{f(x+h)-f(x)}{h}=\dfrac{3(x+h)^2+(x+h)-7-(3x^2+x-7)}{h}$

$=\dfrac{3(x^2+2xh+h^2)+x+h-7-3x^2-x+7}{h}$

$=\dfrac{3x^2+6xh+3h^2+x+h-7-3x^2-x+7}{h}=\dfrac{6xh+3h^2+h}{h}=\dfrac{h(6x+3h+1)}{h}=6x+3h+1$

3. $\dfrac{1}{\sqrt{x+h+3}+\sqrt{x+3}}$ $\dfrac{f(x+h)-f(x)}{h}=\dfrac{\sqrt{x+h+3}-\sqrt{x+3}}{h}$

$=\dfrac{\sqrt{x+h+3}-\sqrt{x+3}}{h}\cdot\dfrac{\sqrt{x+h+3}+\sqrt{x+3}}{\sqrt{x+h+3}+\sqrt{x+3}}=\dfrac{x+h+3-(x+3)}{h(\sqrt{x+h+3}+\sqrt{x+3})}$

$=\dfrac{h}{h(\sqrt{x+h+3}+\sqrt{x+3})}=\dfrac{1}{\sqrt{x+h+3}+\sqrt{x+3}}$

4. $\dfrac{-1}{(x+h+7)(x+7)}$ $\dfrac{f(x+h)-f(x)}{h}=\dfrac{\dfrac{1}{x+h+7}-\dfrac{1}{x+7}}{h}$. For this problem, find a common denominator for the two fractions in the numerator:

$=\dfrac{\dfrac{1}{x+h+7}\cdot\dfrac{x+7}{x+7}-\dfrac{1}{x+7}\cdot\dfrac{x+h+7}{x+h+7}}{h}=\dfrac{\dfrac{x+7}{(x+h+7)(x+7)}-\dfrac{x+h+7}{(x+h+7)(x+7)}}{h}$

$=\dfrac{\dfrac{x+7-x-h-7}{(x+h+7)(x+7)}}{h}=\dfrac{\dfrac{-h}{(x+h+7)(x+7)}}{h}=\dfrac{-h}{(x+h+7)(x+7)}\cdot\dfrac{1}{h}$

In this last step, the complex fraction was simplified by multiplying the numerator by the reciprocal of the denominator.

$=\dfrac{-h}{(x+h+7)(x+7)}\cdot\dfrac{1}{h}=\dfrac{-1}{(x+h+7)(x+7)}$

5. **$3x^2 + 3xh + h^2$** $\dfrac{f(x+h)-f(x)}{h}=\dfrac{(x+h)^3-x^3}{h}=\dfrac{x^3+3x^2h+3xh^2+h^3-x^3}{h}$

$=\dfrac{3x^2h+3xh^2+h^3}{h}=\dfrac{h(3x^2+3xh+h^2)}{h}=3x^2+3xh+h^2$

For more on cubing a binomial, see Chapter 1, "The Basics."

Inverse Functions

A function is a relationship between two variables such that exactly one *output* value exists for every *input* value—only one value in the range for every value in the domain. There are some special functions that also have exactly one *input* value for every *output* value. For instance, the function $y = 3x + 2$ has this property. Look at a chart of some of the coordinates that satisfy the equation.

x	−3	−2	−1	0	1	2	3
y	−7	−4	−1	2	5	8	11

None of the *y* values repeat. No matter what you choose, you can never make that happen. Now look at the function $y = 3x^2 + 2$. This function does **not** have the property that there's only one input value for every output value. The output value 14 has two input values that result in 14. The same is true of the output values 5 and 29 and many more.

x	−3	−2	−1	0	1	2	3
y	29	14	5	2	5	14	29

Functions that have the property in which only one *x* value exists for every *y* value, as well as just one *y* value for every *x* value (that's what makes it a function), are called *one-to-one* functions. In a one-to-one function, if two output values are the same, $f(a) = f(b)$, then it must be that the two input values are equal to one another, $a = b$. The same output can't come from different inputs. What's special about one-to-one functions is that they have *inverses*. Some functions can have inverses very much in the same way that numbers can have inverses.

A number's additive inverse is the number that, when it's added to it, you get 0. The numbers 3 and −3 are additive inverses. A number's multiplicative inverse is the number that, when multiplied by it, gives you 1. The numbers $\frac{2}{3}$ and $\frac{3}{2}$ are multiplicative inverses. Some functions have inverses, too. As we shall see, the composition of a function and its inverse have a very special property.

Showing That Two Functions Are Inverses of One Another

An inverse of a function undoes what the function did. If you perform a function on a number and then perform the inverse function on the result, you're back to the original number. Using the inverse function on an output value tells you what the input value was. Only one-to-one functions can have inverses. An example of a function and its inverse is $f(x) = x^3 − 7$ and $f^{-1}(x) = \sqrt[3]{x + 7}$. The notation $f^{-1}(x)$ doesn't mean to write the reciprocal of the function *f*; it's the standard notation for a function's inverse. Since an inverse of a function takes the function's output and tells you what the original input was, consider the output "1" for the function $f(x) = x^3 − 7$. What value of *x* resulted in that output of 1? Put the 1 into the inverse function, $f^{-1}(1) = \sqrt[3]{1 + 7} = \sqrt[3]{8} = 2$. This tells you that the original input was a 2. Trying it here, $f(2) = 2^3 − 7 = 8 − 7 = 1$. It worked!

Just testing one value like this doesn't prove that one function is the inverse of another. A true test is to evaluate the function and inverse, one in the other, by using the composition of functions. If a function $f(x)$ and $f^{-1}(x)$ are inverses of one another, then $f(x) \circ f^{-1}(x) = f^{-1}(x) \circ f(x) = x$.

Example Problems

These problems show the answers and solutions.

1. Prove that $f^{-1}(x) = \frac{x − 1}{3}$ is the inverse of $f(x) = 3x + 1$.

    answer: They are inverses of one another because $f \circ f^{-1} = f^{-1} \circ f = x$.

$$f(x) \circ f^{-1}(x) = f\left[f^{-1}(x)\right] = 3\left(f^{-1}(x)\right) + 1 = 3\left(\frac{x-1}{3}\right) + 1 = x - 1 + 1 = x$$

$$f^{-1}(x) \circ f(x) = f^{-1}\left[f(x)\right] = \frac{f(x) - 1}{3} = \frac{3x + 1 - 1}{3} = \frac{3x}{3} = x$$

2.   Prove that $f^{-1}(x) = x^3 - 4$ is the inverse of $f(x) = \sqrt[3]{x + 4}$.

    answer: They are inverses of one another, because $f \circ f^{-1} = f^{-1} \circ f = x$.

$$f(x) \circ f^{-1}(x) = f\left[f^{-1}(x)\right] = \sqrt[3]{f^{-1}(x) + 4} = \sqrt[3]{(x^3 - 4) + 4} = \sqrt[3]{x^3} = x$$

$$f^{-1}(x) \circ f(x) = f^{-1}\left[f(x)\right] = \left[f(x)\right]^3 - 4 = \left[\sqrt[3]{x + 4}\right]^3 - 4 = x + 4 - 4 = x$$

Solving for the Inverse of a Function

If a function is one-to-one, then it has an inverse. There's a fairly simple procedure for solving for the inverse of a given one-to-one function. Let $f(x)$ be a function of x.

1. Rewrite the function replacing $f(x)$ with y.

2. Exchange every x with a y and every y with an x.

3. Solve for y.

4. Rewrite the function replacing the y with $f^{-1}(x)$.

Example Problems

These problems show the answers and solutions.

1.   Find the inverse of the function $f(x) = 6x + 7$.

    answer: $f^{-1}(x) = \dfrac{x + 7}{6}$

    Using the procedure outlined previously, first rewrite the problem as $y = 6x + 7$. Now exchange all the x's and y's getting $x = 6y + 7$. Solve for y to get $x - 7 = 6y$ and then $\dfrac{x - 7}{6} = y$. Replace the y with $f^{-1}(x)$ to get $f^{-1}(x) = \dfrac{x + 7}{6}$.

2.   Find the inverse of the function $g(x) = \dfrac{x}{x + 3}$.

    answer: $g^{-1}(x) = \dfrac{3x}{1 - x}$

    Using the procedure given previously, first rewrite the problem as $y = \dfrac{x}{x + 3}$. Now exchange all the x's and y's, resulting in $x = \dfrac{y}{y + 3}$. To solve for y, first multiply each side by the denominator of the fraction. $(y + 3)x = \dfrac{y}{\cancel{y + 3}} \cdot \cancel{(y + 3)}$. Distribute the x on the left: $xy + 3x = y$. Now subtract xy from each side, $3x = y - xy$. Factor out the y on the right side, $3x = y(1 - x)$. Now divide each side by $1 - x$ and change the y to $g^{-1}(x)$: $\dfrac{3x}{1 - x} = y$, $g^{-1}(x) = \dfrac{3x}{1 - x}$.

Work Problems

Use these problems to give yourself additional practice.

1. Prove that $f^{-1}(x) = \dfrac{2x}{3-x}$ is the inverse of $f(x) = \dfrac{3x}{x+2}$.

2. Find the inverse of $f(x) = 4x - 1$.

3. Find the inverse of $f(x) = x^3 - 8$.

4. Find the inverse of $g(x) = \dfrac{1}{x-3}$.

5. Find the inverse of $f(x) = \dfrac{1}{x}$.

Worked Solutions

1. $f(x) \circ f^{-1}(x) = f^{-1}(x) \circ f(x) = x$

$$f(x) \circ f^{-1}(x) = f\left[\frac{2x}{3-x}\right] = \frac{3\left(\frac{2x}{3-x}\right)}{\frac{2x}{3-x}+2} = \frac{\frac{6x}{3-x}}{\frac{2x+6-2x}{3-x}} = \frac{\frac{6x}{3-x}}{\frac{6}{3-x}} = \frac{\cancel{6}x}{\cancel{3-x}} \cdot \frac{\cancel{3-x}}{\cancel{6}} = x$$

$$f^{-1}(x) \circ f(x) = f^{-1}\left[\frac{3x}{x+2}\right] = \frac{2\left(\frac{3x}{x+2}\right)}{\frac{3x+6-3x}{x+2}} = \frac{\frac{6x}{x+2}}{\frac{6}{x+2}} = \frac{\cancel{6}x}{\cancel{x+2}} \cdot \frac{\cancel{x+2}}{\cancel{6}} = x$$

2. $f^{-1}(x) = \dfrac{x+1}{4}$ Exchange the x's and y's to get $x = 4y - 1$. Solve for y by adding 1 to each side and then dividing by 4: $x + 1 = 4y$, $\dfrac{x+1}{4} = y$.

3. $f^{-1}(x) = \sqrt[3]{x+8}$ Exchange the x's and y's to get $x = y^3 - 8$. Solve for y by adding 8 to each side and then taking the cube root of each side: $x + 8 = y^3$, $\sqrt[3]{x+8} = y$.

4. $g^{-1}(x) = \dfrac{1+3x}{x}$ Exchange the x's and y's to get $x = \dfrac{1}{y-3}$. Multiply each side by the denominator and then distribute the x: $x(y - 3) = 1$, $xy - 3x = 1$. Add $3x$ to each side: $xy = 1 + 3x$. Divide each side by x: $y = \dfrac{1+3x}{x}$.

5. $f^{-1}(x) = \dfrac{1}{x}$ This function is its own inverse. Exchange the x's and y's to get $x = \dfrac{1}{y}$. Multiply each side by y and divide each side by x: $xy = 1$, $y = \dfrac{1}{x}$.

Function Transformations

Function transformations are compositions of functions that have particular characteristics. Their effects on a function are predictable, and recognizing these transformations helps you determine quickly and efficiently how a function has changed and what the new, transformed function looks like. The best way to describe these transformations is to consider what they do to the graph of the function. Function transformations can be grouped into three distinct types: slides or translations (up and down and left and right), flips or reflections (over a horizontal or vertical line), and warps (flatter or steeper). The actual graphing of functions using transformations is done in Chapter 6.

Slides

Consider the function $f(x) = x^2$, which is a parabola with its vertex at (0,0), and what this transformation does to it. In each case, C is a positive constant number.

Slide or Translation	Effect on $f(x) = x^2$	Description of Translation
$y = f(x) + C$	$y = x^2 + C$	The graph of $y = x^2$ is raised by C units. The vertex of the new function is at (0,C).
$y = f(x) - C$	$y = x^2 - C$	The graph of $y = x^2$ is lowered by C units. The vertex of the new function is at (0,$-C$).
$y = f(x + C)$	$y = (x + C)^2$	The graph of $y = x^2$ is moved left C units. The vertex of the new function is at ($-C$,0).
$y = f(x - C)$	$y = (x - C)^2$	The graph of $y = x^2$ is moved right C units. The vertex of the new function is at (C,0).

Suppose $C = 10$:

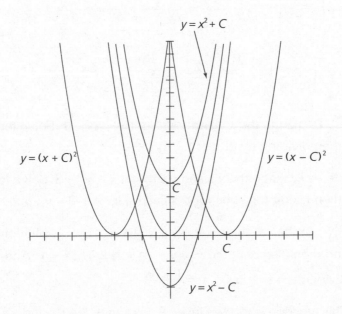

Flips

Consider the function $f(x) = \sqrt{x}$ and what this transformation does to it. In each case, C is a positive constant number.

Flip or Reflection	Effect on $f(x) = \sqrt{x}$	Description of Reflection
$y = -f(x)$	$y = -\sqrt{x}$	The graph of $f(x) = \sqrt{x}$ is reflected over the x-axis.
$y = f(-x)$	$y = \sqrt{-x}$	The graph of $f(x) = \sqrt{x}$ is reflected over the y-axis.

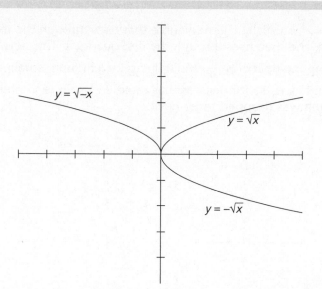

Warps

Consider the function $f(x) = x^2$ and what this transformation does to it. In each case, c is a positive constant number that is greater than 1.

Warp	Effect on $f(x) = x^2$	Description of Warp
$y = c \cdot f(x)$	$y = c \cdot x^2$	The graph of $f(x) = cx^2$ is steeper.
$y = \frac{1}{c} \cdot f(x)$	$y = \frac{1}{c} \cdot x^2$	The graph of $f(x) = \frac{1}{c} x^2$ is flatter.

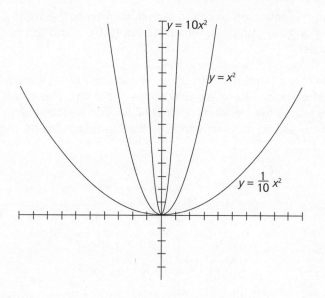

Example Problems

These problems show the answers and solutions.

1. Starting with the basic function $y = x^3$, describe the changes made by the transformations shown with the new function $y = \frac{1}{4} x^3 - 3$.

   answer: The graph is flatter and is lowered by 3 units. The *bend* is now at $(0, -3)$.

The graph of $y = x^3$ is usually a gentle curve that rises through the third quadrant, flattens out at the origin, and then rises through the first quadrant. This new function is flatter because it's being multiplied by $\frac{1}{4}$. Multiplying by a number smaller than 1 has this effect; the y-values don't increase (or decrease) as rapidly. Also, the y-intercept is now at $(0,-3)$, because the graph was dropped three units.

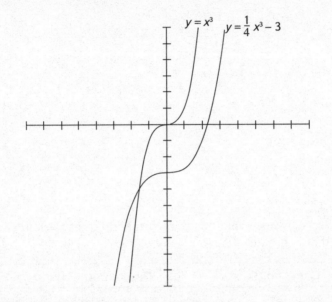

2. Starting with the basic function $y = x^2$, describe the changes made by the transformations shown with the new function $y = -(x - 2)^2 + 5$.

answer: The graph is reflected over the x-axis, moved to the right 2 units and up 5 units. The vertex of the original parabola, which was at $(0,0)$, is moved to $(2,5)$, and the parabola is facing downward.

The graph of $y = x^2$ is a parabola with its vertex at the origin; it rises upward from that vertex. The transformations here are two slides and a flip. The graph slides 2 units to the right and 5 units up, so the vertex is now at $(2,5)$. The negative sign in front flips the whole thing downward so that the vertex is the highest point on the graph.

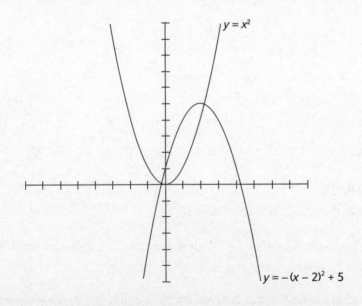

Work Problems

Use these problems to give yourself additional practice.

1. Write the equation of the function that is transformed from $y = x^2$ if the vertex is moved 4 units to the left.

2. Write the equation of the function that is transformed from $y = \sqrt{x}$ if the function becomes steeper by a factor of 5.

3. Write the equation of the function that is transformed from $y = |x|$ if it's moved three units upward and two units to the right.

4. Describe the changes that the transformations on $y = -x^3 + 7$ make to the original function $y = x^3$.

5. Describe the changes that the transformations on $y = \frac{1}{3}(x - 4)^2 - 2$ make to the original function $y = x^2$.

Worked Solutions

1. **$y = (x + 4)^2$**   The graph of this transformation is a slide to the left by 4 units.

2. **$y = 5\sqrt{x}$**   This transformation is a warp. Multiplying by 5 makes the graph of the curve steeper.

3. **$y = |x - 2| + 3$**   There are two slides here.

4. **Up 7 units and flipped over a horizontal line.**   There are two different transformations here. The $-$ sign in front of the variable results in the curve being flipped over a horizontal line. The 7 added to the variable raises the curve by 7 units.

5. **Right 4 and down 2 units; flattened out by a factor of $\frac{1}{3}$.** There are two slides and a warp or flattening of the parabola. The vertex is now at $(4, -2)$, and it rises more slowly because of the factor of $\frac{1}{3}$.

Chapter 4
Polynomials

A polynomial is a smooth curve with a general equation of $y = a_nx^n + a_{n-1}x^{n-1} + \ldots + a_1x^1 + a_0$ where n is a positive integer and the a's are real numbers. The domain of a polynomial is always all real numbers. The number of roots or x-intercepts can be somewhat predicted by the value of n, and other characteristics can be determined by just looking at the equation. This is all covered in this chapter.

Remainder Theorem

Dividing a polynomial by a binomial of the form $x - c$ often results in a remainder. For instance, in the division problem $(x^5 - 13x^3 + 2x^2 - 10x + 5) \div (x - 4)$, there's a remainder of 189; it doesn't divide evenly. To see how to do this using long division or synthetic division, see "Dividing Polynomials" in Chapter 1. The remainder that's obtained has another special value that is useful to know. If the polynomial $f(x)$ is divided by the binomial $x - c$, then the remainder is equal to $f(c)$. What this means is that the remainder is the same as the y-value of the function when $x = c$. Why is this so helpful? Why can't you just substitute the value of c into the equation for $f(x)$? You can do this substitution, of course, and the result usually isn't difficult to compute. But, sometimes, when the powers on the variables are relatively large, doing the synthetic division process to find the remainder is easier than substituting into the equation and finding the values of all those powers. Substitution and synthetic division are two options when computing without a handheld calculator. When a calculator is available, however, it is usually the first choice.

Example Problems

These problems show the answers and solutions.

1. Use the Remainder Theorem to find $f(3)$ if $f(x) = 12x^4 - 33x^3 - 10x^2 - 4x + 17$.

   answer: -4

   Synthetic division is used to find the remainder when $12x^4 - 33x^3 - 10x^2 - 4x + 17$ is divided by $x - 3$.

$$
\begin{array}{r|rrrrr}
3 & 12 & -33 & -10 & -4 & 17 \\
  &   & 36 & 9   & -3 & -21 \\
\hline
  & 12 & 3   & -1 & -7 & -4
\end{array}
$$

   The last number is the remainder, so this is the value of $f(3)$.

This is much easier than doing the powers and division. Compare the preceding to the following:

$$f(3) = 12(3)^4 - 33(3)^3 - 10(3)^2 - 4(3) + 17$$
$$= 12(81) - 33(27) - 10(9) - 12 + 17$$
$$= 972 - 891 - 90 - 12 + 17 = -4$$

2. Use the Remainder Theorem to find $f(-4)$ if $f(x) = 5x^4 + 21x^3 - 6x^2 - 38x + 8$.

    answer: 0

    Using synthetic division,

$$\begin{array}{r|rrrrr} -4 & 5 & 21 & -6 & -38 & 8 \\ & & -20 & -4 & 40 & -8 \\ \hline & 5 & 1 & -10 & 2 & 0 \end{array}$$

    The remainder is 0, which means that when $x = -4$ there's a root or solution of the polynomial. That's where there's an x-intercept when you graph the polynomial.

Work Problems

Use these problems to give yourself additional practice. Use the Remainder Theorem to answer each of the following.

1. Compute $f(3)$ if $f(x) = -6x^5 + 20x^4 - 11x^3 + 16x^2 + 10x - 40$.

2. Compute $f(-8)$ if $f(x) = 10x^4 - 650x^2 - 80x + 2$.

3. Compute $f(2)$ if $f(x) = 15x^4 - 26x^3 + 12x^2 - 38x - 4$.

4. Compute $f(-1)$ if $f(x) = x^8 - 19$.

5. Compute $f\left(\dfrac{1}{3}\right)$ if $f(x) = 6x^4 - 5x^3 - 8x^2 + 9x + 1$.

Worked Solutions

1. **−1**
$$\begin{array}{r|rrrrrr} 3 & -6 & 20 & -11 & 16 & 10 & -40 \\ & & -18 & 6 & -15 & 3 & 39 \\ \hline & -6 & 2 & -5 & 1 & 13 & -1 \end{array}$$

2. **2**
$$\begin{array}{r|rrrrr} -8 & 10 & 0 & -650 & -80 & 2 \\ & & -80 & 640 & 80 & 0 \\ \hline & 10 & -80 & -10 & 0 & 2 \end{array}$$

3. **0**
$$\begin{array}{r|rrrrr} 2 & 15 & -26 & 12 & -38 & -4 \\ & & 30 & 8 & 40 & 4 \\ \hline & 15 & 4 & 20 & 2 & 0 \end{array}$$

4. **−18**
$$
\begin{array}{r|rrrrrrrr}
-1 & 1 & 0 & 0 & 0 & 0 & 0 & 0 & -19 \\
   & -1 & 1 & -1 & 1 & -1 & 1 & -1 & 1 \\
\hline
   & 1 & -1 & 1 & -1 & 1 & -1 & 1 & -1 & -18
\end{array}
$$

5. **3**
$$
\begin{array}{r|rrrrr}
\tfrac{1}{3} & 6 & -5 & -8 & 9 & 1 \\
   & & 2 & -1 & -3 & 2 \\
\hline
   & 6 & -3 & -9 & 6 & 3
\end{array}
$$

Rational Root Theorem and Descartes' Rule of Sign

The Remainder Theorem, discussed in the preceding section, provides a way of determining y-values of a polynomial for given x-values. Some of the more useful y-values are those when y is equal to 0. If y is 0 for a given x, then you have an x-intercept of the graph of the function.

You have several ways of referring to the values that occur when y is equal to 0: x-intercept, solution, root, and zero. They're often used interchangeably. Even though the "answer" is correct no matter which wording you use, it's best to use the correct ones. x-intercepts are really just that—where the graph of the function crosses the x-axis. Solutions and roots are answers to problems in which you set something equal to 0 and solve for the value or values that make it so. Zero just means find out what *makes* the expression equal to 0.

To find these intercepts/roots/zeros, you can keep trying different numbers in synthetic division for the Remainder Theorem, or you can make a plan, eliminate unnecessary attempts, and adjust your approach as you get more information. The plan is determined from the Rational Root Theorem and Descartes' Rule of Sign. The adjustments are made as roots are found.

Rational Root Theorem

If the polynomial $f(x)$ has any rational roots, they will all be numbers of the form $\frac{c}{a}$ where a is a factor of the lead coefficient, a_n, and a is a factor of the constant term, a_0. $\frac{c}{a} = \frac{\text{factor of } a_0}{\text{factor of } a_n}$.

Descartes' Rule of Sign

To determine the *maximum* number of positive real roots that a polynomial $f(x)$ can have, count the number of sign changes between the terms of the polynomial (written in decreasing powers). To determine the *maximum* number of negative real roots that a polynomial can have, count the number of sign changes between the terms of $f(-x)$. The maximum won't necessarily be how many roots there are. Sometimes, it's fewer. But, in both cases, if the polynomial doesn't have the maximum number of a type of root, then there are two fewer (or four, or six, and so on) than that. The number decreases by twos.

If you need to review what rational and real numbers are, refer to Chapter 1 of this book.

Example Problems

These problems show the answers and solutions.

1. Find the zeros/solutions/roots of $f(x) = x^4 + 8x^3 + 11x^2 - 32x - 60$.

    answer: $2, -2, -3, -5$

    This is a fourth degree polynomial, so there could be as many as four different solutions to $f(x) = 0$. According to the Rational Root Theorem, any rational root has to be of the form $\frac{\text{factor of } 60}{\text{factor of } 1}$. So the possibilities are $\pm1, \pm2, \pm3, \pm4, \pm5, \pm6, \pm10, \pm12, \pm15, \pm20, \pm30,$ ±60. That's a lot of numbers to have to try. Descartes' Rule of Sign will help narrow down the search for zeros. The function $f(x) = x^4 + 8x^3 + 11x^2 - 32x - 60$ has just one change in sign—from positive to negative. So there's only one positive real root. Replacing all the x's with $-x$'s, $f(-x) = x^4 - 8x^3 + 11x^2 + 32x - 60$. This has three changes of sign: from positive to negative, negative to positive, and positive to negative. This means that there are at most three negative real roots. If there aren't three, then there is one.

    The next part involves clever guessing as to what the solutions are, coupled with synthetic division to find out when the remainder is 0.

    Since there are probably more negative roots than positive, first try -1.

    $$
    \begin{array}{r|rrrrr}
    -1 & 1 & 8 & 11 & -32 & -60 \\
       &   & -1 & -7 & -4 & 36 \\
    \hline
       & 1 & 7 & 4 & -36 & -24
    \end{array}
    $$

    That didn't work, so try -2.

    $$
    \begin{array}{r|rrrrr}
    -2 & 1 & 8 & 11 & -32 & -60 \\
       &   & -2 & -12 & 2 & 60 \\
    \hline
       & 1 & 6 & -1 & -30 & 0
    \end{array}
    $$

    That one did work. The number -2 is a root or solution. This synthetic division is a shortcut to long division, and the numbers across the bottom are the quotient. Dividing by $x + 2$ leaves a factor that's the same as the quotient. Take advantage of this and just use the quotient, which is the list of numbers across the bottom of the synthetic division, to try the next number. Try another negative number, -3.

    $$
    \begin{array}{r|rrrr}
    -3 & 1 & 6 & -1 & -30 \\
       &   & -3 & -9 & 30 \\
    \hline
       & 1 & 3 & -10 & 0
    \end{array}
    $$

    That's another solution! Now you're down to three terms in the quotient, which corresponds to the trinomial $x^2 + 3x - 10$. This factors into $(x + 5)(x - 2)$. Setting that equal to 0, $(x + 5)(x - 2) = 0$, the two remaining roots are -5 and 2. The four roots/solutions/zeros are $-2, -3, -5, 2$. Just as predicted, there are three negative roots and one positive.

2. Find the zeros/solutions/roots of $f(x) = 3x^3 - 35x^2 + 98x - 80$.

answer: $\frac{5}{3}$, 2, 8

This is a cubic polynomial, so there could be as many as three different solutions. First, to look at all of the possibilities, $\frac{\text{factor of } 80}{\text{factor of } 3}$, the factors of 3 are 1 and 3, so first list all of the factors of 80 divided by 1. $\pm1, \pm2, \pm4, \pm5, \pm8, \pm10, \pm16, \pm20, \pm40, \pm80$. Then go back and divide all of them by 3: $\pm\frac{1}{3}, \pm\frac{2}{3}, \pm\frac{4}{3}, \pm\frac{5}{3}, \pm\frac{8}{3}, \pm\frac{10}{3}, \pm\frac{16}{3}, \pm\frac{20}{3}, \pm\frac{40}{3}, \pm\frac{80}{3}$.

Again, this is a lot of choices. To narrow them down, hopefully, look at the signs. In the function $f(x)$, the sign changes three times, so there are either three positive real roots or there's only one. Checking for negative real roots, you replace the x with $-x$, so $f(-x) = -3x^3 - 35x^2 - 98x - 80$. The sign doesn't change at all, so there are no negative real roots. Concentrate on trying positive roots, and the first one to try is $x = 2$. (Yes, I peeked.)

$$
\begin{array}{r|rrrr}
2 & 3 & -35 & 98 & -80 \\
  &   & 6 & -58 & 80 \\
\hline
  & 3 & -29 & 40 & 0
\end{array}
$$

The 2 worked, and there are just three terms left. They represent the trinomial $3x^2 - 29x + 40$. This factors into $(3x - 5)(x - 8)$. Setting it equal to 0, the solutions are $\frac{5}{3}$ and 8. The three positive solutions are 2, $\frac{5}{3}$, and 8. Notice that they're all in the list.

Work Problems

Use these problems to give yourself additional practice. Find the roots/solutions/zeros of the following.

1. $f(x) = x^3 - 4x^2 + x + 6$

2. $f(x) = 2x^3 + x^2 - 50x - 25$

3. $f(x) = x^4 - 9x^3 + 9x^2 + 49x + 30$

4. $f(x) = 8x^3 - 12x^2 + 6x - 1$

5. $f(x) = 9x^4 - 37x^2 + 4$

Worked Solutions

1. **−1, 2, 3**

The possible rational solutions, according to the Rational Root Theorem, are $\pm1, \pm2, \pm3, \pm6$. There are two sign changes in $f(x)$, so there are a maximum of two positive real roots. $f(-x) = -x^3 - 4x^2 - x + 6$ has one sign change, so there's one negative real root. If a choice for a root is 2, the synthetic division would be

$$
\begin{array}{r|rrrr}
2 & 1 & -4 & 1 & 6 \\
  &   & 2 & -4 & -6 \\
\hline
  & 1 & -2 & -3 & 0
\end{array}
$$

The 2 is a solution. The quotient represents the trinomial $x^2 - 2x - 3$, which can be factored into $(x - 3)(x + 1)$. Set that equal to 0, $x = 3, -1$. This isn't the only way that this could be solved. It all depends on what you guess the solutions to be.

2. **5, -5, $-\dfrac{1}{2}$**

   The possible rational solutions, according to the Rational Root Theorem, are $\pm 1, \pm 5, \pm 25,$ $\pm\dfrac{1}{2}, \pm\dfrac{5}{2}, \pm\dfrac{25}{2}$. There is one sign change in $f(x)$, so there is one positive real root. $f(-x) =$ $-2x^3 + x^2 + 50x - 25$ has two sign changes, so there's a maximum of two negative real roots. If a choice for a root is -5, the synthetic division would be

$$
\begin{array}{r|rrrr}
-5 & 2 & 1 & -50 & -25 \\
   &   & -10 & 45 & 25 \\
\hline
   & 2 & -9 & -5 & 0
\end{array}
$$

   The -5 is a solution. The quotient represents the trinomial $2x^2 - 9x - 5$, which can be factored into $(2x + 1)(x - 5)$. Setting that equal to 0, the two solutions are $-\dfrac{1}{2}, 5$.

3. **5, 6, -1**

   The possible rational solutions, according to the Rational Root Theorem, are $\pm 1, \pm 2, \pm 3,$ $\pm 5, \pm 6, \pm 10, \pm 15, \pm 30$. There are two sign changes in $f(x)$, so there is a maximum of two positive real roots. $f(-x) = x^4 + 9x^3 + 9x^2 - 49x + 30$ has two sign changes, so there's a maximum of two negative real roots. If a choice for a root is 5, the synthetic division would be

$$
\begin{array}{r|rrrrr}
5 & 1 & -9 & 9 & 49 & 30 \\
  &   & 5 & -20 & -55 & -30 \\
\hline
  & 1 & -4 & -11 & -6 & 0
\end{array}
$$

   The 5 is a solution. The quotient represents a third degree polynomial with no obvious way of factoring, so another application of synthetic division will be used on it. The list of possible rational roots has changed. The quotient represents the polynomial $x^3 - 4x^2 - 11x - 6$. Using the Rational Root Theorem, the choices are now $\pm 1, \pm 2, \pm 3, \pm 6$. All the numbers larger than 6 have been eliminated. Trying 6,

$$
\begin{array}{r|rrrr}
6 & 1 & -4 & -11 & -6 \\
  &   & 6 & 12 & 6 \\
\hline
  & 1 & 2 & 1 & 0
\end{array}
$$

   The 6 is a root. The quotient represents $x^2 + 2x + 1$, which factors into $(x + 1)(x + 1) =$ $(x + 1)^2$. This last part of the equation tells you that there's a double root. The solution -1 occurs twice.

4. **$\dfrac{1}{2}$**

   In this problem, there's a triple root. There's only one value for the solutions. The Rational Root Theorem provides the choices $\pm 1, \pm\dfrac{1}{2}, \pm\dfrac{1}{4}, \pm\dfrac{1}{8}$. There are three sign changes in $f(x)$, so there's a maximum of three positive real roots. $f(-x) = -8x^3 - 12x^2 - 6x - 1$ has no change in sign, so there are no negative real roots. That helps to narrow the search. Most people would probably start with the 1 as a possible solution. That isn't going to work.

Using the $\frac{1}{2}$,

$$\frac{1}{2}\big|\ \begin{array}{rrrr} 8 & -12 & 6 & -1 \\ & 4 & -4 & 1 \\ \hline 8 & -8 & 2 & 0 \end{array}$$

This leaves a quotient represented by $8x^2 - 8x + 2$, which factors as $2(4x^2 - 4x + 1) = 2(2x - 1)^2$. Setting that equal to 0 results in the other two solutions.

5. **$2, -2, \dfrac{1}{3}, -\dfrac{1}{3}$**

The possible rational solutions, according to the Rational Root Theorem, are ± 1, ± 2, ± 4, $\pm\frac{1}{3}$, $\pm\frac{2}{3}$, $\pm\frac{4}{3}$, $\pm\frac{1}{9}$, $\pm\frac{2}{9}$, $\pm\frac{4}{9}$. There are two sign changes in $f(x)$, so there is a maximum of two positive real roots. $f(-x) = -9x^4 - 37x^2 + 4$ has two sign changes, so there's a maximum of two negative real roots. If a choice for a root is 2, the synthetic division would be

$$2\big|\ \begin{array}{rrrrr} 9 & 0 & -37 & 0 & 4 \\ & 18 & 36 & -2 & -4 \\ \hline 9 & 18 & -1 & -2 & 0 \end{array}$$

The 2 is a solution. The polynomial that's left can be factored by grouping. Also, another application of synthetic division can be used. Factor by grouping, $9x^3 + 18x^2 - x - 2 = 9x^2(x + 2) - 1(x + 2) = (x + 2)(9x^2 - 1) = (x + 2)(3x - 1)(3x + 1)$. Setting the factored form equal to 0, the other solutions are -2, $\frac{1}{3}$, $-\frac{1}{3}$. Looking back at the original problem, you may have noticed that the fourth-degree polynomial could be factored, because it's a quadratic-like trinomial. Refer to "Factoring Trinomials" in Chapter 2 for more on this.

Upper and Lower Bounds

Searching for roots or solutions of polynomials can be made easier with the Rational Root Theorem and Descartes' Rule of Sign, found in the previous section of this chapter. There's another aid to this search. You determine *bounds* or values that the solutions either can't exceed or can't go below. If, for instance, you list all of the possible rational roots as being ± 1, ± 2, ± 3, ± 5, ± 6, ± 10, ± 15, ± 30, and you then determine that an upper bound for the solution is 10, then you can ignore the 15 and the 30 in the list of choices. If you also determine that a lower bound is -2, then you can discard the -3, -5, -6, -10, -15, -30. This process helps narrow down the search and save time.

How do you determine those bounds? First, be sure that your polynomial starts with a positive term. If you have $f(x) = -12x^3 + 8x^2 + 3x - 2$ and want to find the zeros or solutions to $-12x^3 + 8x^2 + 3x - 2 = 0$, then first multiply each side by -1 to get $12x^3 - 8x^2 - 3x + 2 = 0$. This will not affect your solutions. The Rational Root Theorem says that the possible rational roots are ± 1, ± 2, $\pm\frac{1}{2}$, $\pm\frac{1}{3}$, $\pm\frac{1}{4}$, $\pm\frac{1}{6}$, $\pm\frac{1}{12}$, $\pm\frac{2}{3}$. Now consider the bounds.

Upper Bound: If you perform synthetic division on the coefficients using a *positive* number, and the quotient (bottom row of numbers) contains all positive numbers or 0's, then the number you're dividing by is an upper bound. There's no solution larger than that number.

Lower Bound: If you perform synthetic division on the coefficients using a *negative* number, and the quotient (bottom row of numbers) has alternating $+$ and $-$ signs, then the number you're dividing with is a lower bound. There's no root smaller. If a 0 appears in that row, then it can be assigned a $+$ or $-$ sign to make the signs alternate.

Example Problems

These problems show the answers and solutions.

1. Show that the numbers 1 and -1 are upper and lower bounds, respectively, of the zeros of $f(x) = 12x^3 - 8x^2 - 3x + 2$.

$$\begin{array}{r|rrrr} 1 & 12 & -8 & -3 & 2 \\ & & 12 & 4 & 1 \\ \hline & 12 & 4 & 1 & 3 \end{array}$$

The last row has all positive numbers, so 1 is an upper bound.

$$\begin{array}{r|rrrr} -1 & 12 & -8 & -3 & 2 \\ & & -12 & 20 & -17 \\ \hline & 12 & -20 & 17 & -15 \end{array}$$

The last row has numbers that alternate between $+$ and $-$, so -1 is a lower bound.

2. Find an upper bound and lower bound for the zeros of $f(x) = x^4 + 2x^3 - 19x^2 + 28x - 12$.

    answer: 4 and -6

    The Rational Root Theorem provides possible rational roots as being ±1, ±2, ±3, ±4, ±6, ±12.

$$\begin{array}{r|rrrrr} 4 & 1 & 2 & -19 & 28 & -12 \\ & & 4 & 24 & 20 & 192 \\ \hline & 1 & 6 & 5 & 48 & 180 \end{array}$$

    4 is an upper bound, because the last row of numbers is all $+$.

$$\begin{array}{r|rrrrr} -6 & 1 & 2 & -19 & 28 & -12 \\ & & -6 & 24 & -30 & 12 \\ \hline & 1 & -4 & 5 & -2 & 0 \end{array}$$

    -6 is a lower bound, because the last row alternates in signs, and the 0 can be considered to be $+$ to complete the pattern.

Work Problems

Use these problems to give yourself additional practice. Determine whether or not the given values are upper and lower bounds for the zeros of the polynomials.

1. $f(x) = x^3 - 12x^2 + 44x - 48$; upper, 12; lower, -1

2. $g(x) = x^3 + 11x^2 + 20x - 32$; upper, 2; lower, -3

3. $h(x) = x^4 + 3x^3 - 5x^2 - 3x + 4$; upper, 2; lower, -2

4. $k(x) = 9x^4 + 15x^3 - 26x^2 - 64x - 32$; upper 1; lower, -1

5. $p(x) = x^4 - 29x^2 + 100$; upper, 2; lower, -10

Worked Solutions

1. **Yes, 12 is an upper bound; yes, −1 is a lower bound.**

$$
\begin{array}{r|rrrr}
12 & 1 & -12 & 44 & -48 \\
   &   & 12 & 0 & 528 \\
\hline
   & 1 & 0 & 44 & 480
\end{array}
\qquad
\begin{array}{r|rrrr}
-1 & 1 & -12 & 44 & -48 \\
   &   & -1 & 13 & -57 \\
\hline
   & 1 & -13 & 57 & -105
\end{array}
$$

2. **Yes, 2 is an upper bound; no, −3 is not a lower bound.**

$$
\begin{array}{r|rrrr}
2 & 1 & 11 & 20 & -32 \\
  &   & 2 & 26 & 92 \\
\hline
  & 1 & 13 & 46 & 60
\end{array}
\qquad
\begin{array}{r|rrrr}
-3 & 1 & 11 & 20 & -32 \\
   &   & -3 & -24 & 12 \\
\hline
   & 1 & 8 & -4 & -20
\end{array}
$$

3. **Yes, 2 is an upper bound; no, −2 is not a lower bound.**

$$
\begin{array}{r|rrrrr}
2 & 1 & 3 & -5 & -3 & 4 \\
  &   & 2 & 10 & 10 & 14 \\
\hline
  & 1 & 5 & 5 & 7 & 18
\end{array}
\qquad
\begin{array}{r|rrrrr}
-2 & 1 & 3 & -5 & -3 & 4 \\
   &   & -2 & -2 & 14 & -22 \\
\hline
   & 1 & 1 & -7 & 11 & 18
\end{array}
$$

4. **No, 1 is not an upper bound; no, −1 is not a lower bound.**

$$
\begin{array}{r|rrrrr}
1 & 9 & 15 & -26 & -64 & -32 \\
  &   & 9 & 24 & -2 & -66 \\
\hline
  & 9 & 24 & -2 & -66 & -98
\end{array}
\qquad
\begin{array}{r|rrrrr}
-1 & 9 & 15 & -26 & -64 & -32 \\
   &   & -9 & -6 & 32 & 32 \\
\hline
   & 9 & 6 & -32 & -32 & 0
\end{array}
$$

5. **No, 2 is not an upper bound; yes, −10 is a lower bound.**

$$
\begin{array}{r|rrrrr}
2 & 1 & 0 & -29 & 0 & 100 \\
  &   & 2 & 4 & -50 & -100 \\
\hline
  & 1 & 2 & -25 & -50 & 0
\end{array}
\qquad
\begin{array}{r|rrrrr}
-10 & 1 & 0 & -29 & 0 & 100 \\
    &   & -10 & 100 & -710 & 7100 \\
\hline
    & 1 & -10 & 71 & -710 & 7200
\end{array}
$$

Using a Graphing Calculator to Graph Lines and Polynomials

Graphing calculators are very powerful tools. They can increase accuracy and efficiency and allow you to spend your time on the more important and interesting aspects of the work. But, with all this technological power available, there's no substitute for thinking and planning and knowing what to expect. Graphing calculators can provide you with wonderful results, or they can provide you with absolute nonsense.

Lines are predictable—you know what you expect to see when you graph them. Polynomials are also relatively predictable—you know that their domains are all real numbers and that their y-values can get very large or small. If you don't tell your graphing calculator where to look for these lines or polynomials, you can miss their graphs completely. Sometimes, graphing a function with a calculator can be like looking for a lifeboat in the middle of the ocean. You know what you're looking for, but you have to have some clues to find it. When you know some of the characteristics of these graphs, it's like having a global positioning system—you have clues as to where to look.

A main key in using graphing calculators to graph lines and polynomials is to make use of the intercepts—both x-intercepts and y-intercepts. After you've determined where they are, you can use the *autofit* or *autoscale* capability of your calculator to make the appropriate settings as to how high or how low the graph will go.

Example Problems

These problems show the answers and solutions.

1. Graph the line $40x - 13y = 520$

    answer:

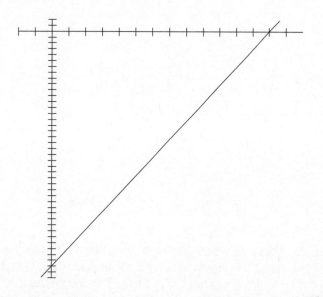

First, rewrite the equation of the line in slope-intercept form, $y = \frac{40}{13}x - 40$, and enter this into the y_1 position of the y-menu in your calculator. Most graphing calculators look at that first entry when doing *autofit* or *autoscale*. Also, graphing calculators have a standard setting that goes from -10 to 10 both horizontally and vertically. If you graph this line using the standard setting, you'll see nothing but a blank screen. From its equation, you can see that the y-intercept of this line is -40. That's why it doesn't show up in a standard setting. Find the x-intercept of the line by going back to the original equation, letting y equal 0 and solving for x. $40x - 13y = 520$, $40x = 520$, $x = 13$. The x-intercept is at $(13,0)$. The two intercepts of this line are at $(0,-40)$ and $(13,0)$. You want to include these two intercepts in your graph, so go to the *window* or *range* setting of your calculator and choose your x-minimum to be -2 and the x-maximum to be 15. This gives you 2 units to the left of 0 and 2 units to the right of 13. The 0 and 13 are the x-values of the two intercepts. Now use *autofit* or *autoscale* to get your graph. Notice that the scale on the x-axis is different from that on the y-axis. This happens automatically so that the whole picture can be included.

2.   Graph $y = x^4 - x^3 - 114x^2 + 4x + 440$

answer:

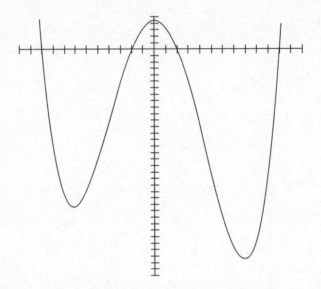

Find the *x*-intercepts. These are the roots/solutions/zeros. Go to the section on "Rational Root Theorem" if you need to review this. The roots are: $-10, -2, 2, 11$. This means that the *x*-intercepts are $(-10,0), (-2,0), (2,0), (11,0)$. The *y*-intercept is 440. This graph actually goes pretty high and low between the intercepts. Go to *window* or *range* and set the *x*-minimum at -12 and the *x*-maximum at 13. The -12 is 2 units smaller than the lowest *x*-intercept, and the 13 is 2 units larger than the highest *x*-intercept. Use *autofit* or *autoscale* to get your graph. Sometimes, you'll want to use a wider window to get a more complete picture. If you want to adjust it a bit, change the *x*-minimum and *x*-maximum, and then just push the *graph* button. When the numbers get really large, you'll lose much of the detail close to the axes. The trade-off is that you do see the big picture.

Work Problems

Use these problems to give yourself additional practice.

1.   Graph the line $25x + 18y = 900$.

2.   Graph the parabola $y = x^2 - 9x - 10$.

3.   Graph the polynomial $y = 100x - x^3$.

4.   Graph $y = -3x(x - 15)(x + 17)$.

5.   Graph $y = 64x^4 - 20x^2 + 1$.

Worked Solutions

1.

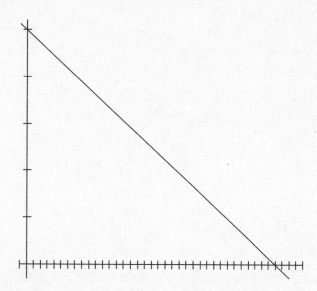

The intercepts are (0,50) and (36,0). Put $y = -\frac{25}{18}x + 50$ in the y_1 position of your y-menu. Set the x-minimum and x-maximum at 0 and 36, respectively. Use *autofit* to see the graph.

2.

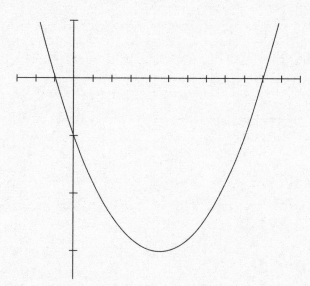

The x-intercepts are -1 and 10. Put $y = x^2 - 9x - 10$ in the y_1 position of your y-menu. Set the x-minimum and x-maximum at -3 and 12, respectively. Use *autofit* to see the graph.

3.

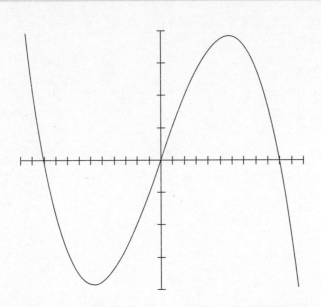

The x-intercepts are −10, 0, and 10. Put $y = 100x − x^3$ in the y_1 position of your y-menu. Set the x-minimum and x-maximum at −12 and 12, respectively. Use *autofit* to see the graph.

4.

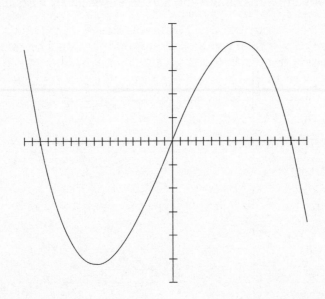

The x-intercepts are −17, 0, and 15. Put $y = −3x(x − 15)(x + 17)$ in the y_1 position of your y-menu. Set the x-minimum and x-maximum at −19 and 17, respectively. Use *autofit* to see the graph.

5.

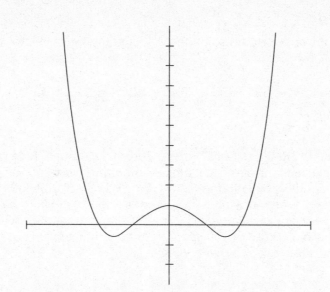

This factors $(4x^2 - 1)(16x^2 - 1) = (2x - 1)(2x + 1)(4x - 1)(4x + 1)$. Setting this equal to 0, the x-intercepts are $-\frac{1}{2}, -\frac{1}{4}, \frac{1}{4}, \frac{1}{2}$. In this case, there's the opposite problem of having everything spread out so far. The graph of this curve has all the detail really close to the origin. Put $y = 64x^4 - 20x^2 + 1$ in the y_1 position of your y-menu. Set the x-minimum and x-maximum at -1 and 1, respectively. Use *autofit* to see the graph. Even this setting doesn't give you much detail. If you look at the window, you'll see that the *autofit* gives a pretty large value for the y-maximum. Change that maximum to a 10, push *graph*, and you'll see a nicer picture.

Solving Equations and Inequalities Using Graphing

Polynomials have solutions, x-intercepts, or roots that are actually the x-values or input values that make the expression equal to 0. A major part of graphing polynomials and determining other things about them, such as local maximum values or local minimum values, is to first find those intercepts or zeros.

A *local maximum* or *local minimum* is found at a point that's either higher than any point close to it or lower than any point close to it. A polynomial can have a global maximum (or global minimum) if there is no value higher (or lower) than it on the entire graph. These occur when the highest power is even. Polynomials whose highest power is an odd number won't have a global maximum or minimum, because their output values will go to positive or negative infinity as x gets very large or very small. Polynomials whose highest power is an even number will either have a highest or lowest value, depending on whether the coefficient on that highest power is positive or negative.

The easiest way to show all of these concepts is to give some examples.

Example Problems

These problems show the answers and solutions. Use a graphing calculator to help you solve these problems.

1. Solve the equation $3x^4 + 5x^3 - 5x^3 - 50x^2 - 80x + 32 = 0$.

    answer: $-4, -2, \frac{1}{3}, 4$

    This can be solved using the Rational Root Theorem, Descartes' Rule of Sign, and bounds, discussed in earlier sections of this chapter. Another big help to finding the zeros is to graph the corresponding equation $y = 3x^4 + 5x^3 - 5x^3 - 50x^2 - 80x + 32$ to see whether you can tell where the intercepts seem to be—or close to being. Putting this equation into a graphing calculator and using the standard setting where the x and y values both go from -10 to 10, you seem to get nearly vertical lines that appear to cross at $-4, -2, 4$, and somewhere between 0 and 1.

    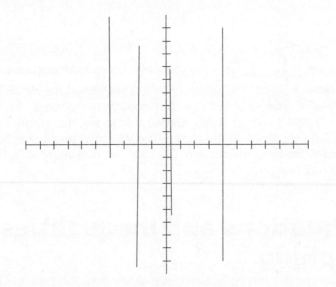

    The Rational Root Theorem provides a list of possible solutions: ± 1, ± 2, ± 4, ± 8, ± 16, ± 32, $\pm\frac{1}{3}$, $\pm\frac{2}{3}$, $\pm\frac{4}{3}$, $\pm\frac{8}{3}$, $\pm\frac{16}{3}$, $\pm\frac{32}{3}$. The numbers -4, -2, and 4 are on the list, and if you check them out, they are solutions. The root between 0 and 1 could be $\frac{1}{3}$ or $\frac{2}{3}$.

    $$\begin{array}{r|rrrrr}
    \frac{1}{3} & 3 & 5 & -50 & -80 & 32 \\
    & & 1 & 2 & -16 & -32 \\
    \hline
    & 3 & 6 & -48 & -96 & 0
    \end{array}$$

    As you see, it's the $\frac{1}{3}$ that works.

2. Solve $x^5 - 4x^4 - 4x^3 + 22x^2 + 3x - 18 \geq 0$.

    answer: $-2 \leq x \leq -1, x \geq 1$

    First, put the corresponding equation, $y = x^5 - 4x^4 - 4x^3 + 22x^2 + 3x - 18$ in your calculator and graph it using the standard setting. The graph appears to be straight lines and a V shape. To get a better picture, find the zeros and use *autofit* or *autoscale*.

The zeros appear to be -2, -1, 1, 3. The list of roots, using the Rational Root Theorem, are ± 1, ± 2, ± 3, ± 6, ± 9, ± 18. Using synthetic division to test these,

$$
\begin{array}{r|rrrrrr}
-2 & 1 & -4 & -4 & 22 & 3 & -18 \\
   &   & -2 & 12 & -16 & -12 & 18 \\
\hline
-1 & 1 & -6 & 8 & 6 & -9 & 0 \\
   &   & -1 & 7 & -15 & 9 & \\
\hline
1 & 1 & -7 & 15 & -9 & 0 & \\
  &   & 1 & -6 & 9 & & \\
\hline
3 & 1 & -6 & 9 & 0 & & \\
  &   & 3 & -9 & & & \\
\hline
3 & 1 & -3 & 0 & & & \\
  &   & 3 & & & & \\
\hline
  & 1 & 0 & & & & \\
\end{array}
$$

The roots that were identified from the graph, -2, -1, 1, 3, are confirmed with successive applications of synthetic division. The double root at 3 appears here, too. When a zero or root occurs an even number of times—twice, four times, and so on—then the graph at that zero appears to be a touch and go. The graph of the curve touches at that point, but it doesn't cross the x-axis to the other side. For this reason, graphing calculators usually can't find that zero or root with their built-in root finding program, because there's no sign change.

Now use *autofit*, going to the window or range and setting the x-minimum at -3 and the x-maximum at 5.

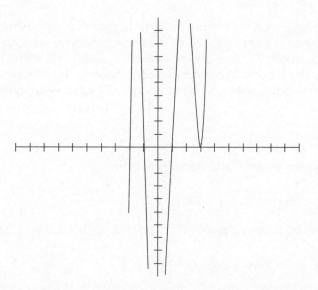

This doesn't give you much definition near the x-axis, so go back to the window or range and change the y minimum to -50. Push the *graph* button. Now you're ready to go back to the original question, which asks when the expression is positive, $x^5 - 4x^4 - 4x^3 + 22x^2 + 3x - 18 \geq 0$. The graph of the curve intersects the x-axis at -2 and -1 and is above the x-axis or positive between those two values. It also intersects the x-axis at 1 and 3 and is above the x-axis or positive between those values. The curve is also above the axis when it's to the right of 3. So, the answer to the original question is that the expression is 0 or positive when $-2 \leq x \leq -1$, $x \geq 1$, which is written as $[-2, -1] \cup [1, \infty]$ in interval notation.

3. Estimate where there's a local maximum y value and local minimum y value for
   $y = x^3 - 2x^2 - 29x + 30$.

   answer: maximum at about $(-2.5, 74)$; minimum at about $(3.7, -54)$

   Putting the equation into a graphing calculator and using the standard setting to look at the graph, there appear to be zeros or roots at -5, 1, and 6. Now set the window or range with an x-minimum of -6 and a x-maximum of 7.

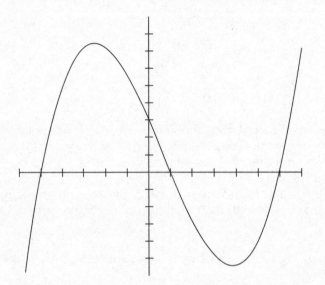

   There appears to be a local maximum (a point higher than any around it) when x is somewhere between -3 and -2. Using the *trace* capability on your calculator, the highest value seems to be about 74. This occurs when x is about -2.5. The local minimum (a point lower than any around it) appears to occur when x is between 3 and 4. Using the *trace* again, the low point seems to be at about -54 when x is about 3.7.

Work Problems

Use these problems to give yourself additional practice.

1. Solve $x^4 - 10x^3 + 35x^2 - 50x + 24 = 0$.

2. Solve $x^4 - 33x^2 + 8x + 240 \leq 0$.

3. Solve $x^4 - 7x^3 + 2x^2 + 64x - 96 > 0$.

4. Solve $x^6 - x^5 - 129x^4 + 133x^3 + 968x^2 - 1452x = 0$.

5. Find the local maximum and local minimum values for $y = x^4 - x^2 - 6$.

Worked Solutions

1. **$x = 1, 2, 3, 4$**   Look at the graph of $y = x^4 - 10x^3 + 35x^2 - 50x + 24$ in the standard window. The graph has vertical lines that appear to cross at 1, 2, 3, and 4. Test these values using synthetic division.

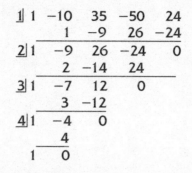

$$
\begin{array}{r|rrrrr}
1 & 1 & -10 & 35 & -50 & 24 \\
  &   & 1 & -9 & 26 & -24 \\
\hline
2 & 1 & -9 & 26 & -24 & 0 \\
  &   & 2 & -14 & 24 & \\
\hline
3 & 1 & -7 & 12 & 0 & \\
  &   & 3 & -12 & & \\
\hline
4 & 1 & -4 & 0 & & \\
  &   & 4 & & & \\
\hline
  & 1 & 0 & & &
\end{array}
$$

2. **$-5 \le x \le -3$ or $x = 4$**   Look at the graph of $y = x^4 - 33x^2 + 8x + 240$ in your calculator using the standard setup for starters. There appears to be zeros at $-5, -3$, and 4. Adjust the range/window so that the minimum x is -6 and the maximum is 5. Then use *autofit* to get a new picture. The graph of the function is above the x-axis for all values less than -3, between 2 and 4, and then from 4 on up. There's a "touch and go" at 4; the y-value is 0 there, but it doesn't cross the x-axis to the other side. The only place that the graph goes beneath the x-axis is between -5 and -3. This satisfies the original problem, looking for when the value of the expression is negative or 0.

3. **$x < -3, 2 < x < 4, x > 4$**   Look at the graph of $y = x^4 - 7x^3 + 2x^2 + 64x - 96$ in your calculator using the standard setup for starters. The graph appears to be above the x-axis for values lower than -3, below the x-axis between -3 and 2, and then back above the x-axis for values greater than 2. There's a "touch and go" at 4. This time, the value 4 can't be used in the solution. The problem asks for positive values, only. That's why the answer splits up at 4. Another way of writing the solution is in interval notation: $(-\infty, 3) \cup (2, 4) \cup (4, \infty)$.

4. **$x = -11, -3, 0, 2, 11$**   Look at the graph of $y = x^6 - x^5 - 129x^4 + 133x^3 + 968x^2 - 1452x$ using the standard setting. There appear to be vertical lines crossing at -3, 0, and 2. Using synthetic division, these solutions can be verified and the other three roots discovered. But, before using synthetic division, the factor x has to be removed. $x^6 - x^5 - 129x^4 + 133x^3 + 968x^2 - 1452x = x(x^5 - x^4 - 129x^3 + 133x^2 + 968x - 1452$. Now synthetic division can be performed on the second factor.

$$
\begin{array}{r|rrrrrr}
-3 & 1 & -1 & -129 & 133 & 968 & -1452 \\
   &   & -3 & 12 & 351 & -1452 & 1452 \\
\hline
2 & 1 & -4 & -117 & 484 & -484 & 0 \\
  &   & 2 & -4 & -242 & 484 & \\
\hline
  & 1 & -2 & -121 & 242 & 0 &
\end{array}
$$

The quotient (last line) represents the polynomial $x^3 - 2x^2 - 121x + 242$. This can be factored using grouping into $x^2(x - 2) - 121(x - 2) = (x - 2)(x^2 - 121) = (x - 2)(x + 11)(x - 11)$. There's another 2 (a double root at 2) and zeros of 11 and -11.

5. **local maximum at about $(0, -6)$; local minimum values at about $(-.7, -6.25)$ and $(.7, -6.25)$**   Look at the graph of the function using the standard setting. You don't get much definition in the lower portion, so change the window so that the x values go from

−3 to 3 and the *y* values go from −10 to 1. There appears to be a local maximum on the *y*-axis and two local minimum values on either side of it. Use trace to approximate where these values are occurring.

Chapter 5
Exponential and Logarithmic Functions

Exponential and logarithmic functions make up a very important part of the list of functions that model real-life situations. Many things increase or decrease *exponentially*. The money in an account earning interest grows *exponentially*. The number of amoebas in a Petri dish can grow *exponentially*. The amount of carbon-14 in a dead plant will decrease *exponentially*. Logarithmic functions appear in models involving the pH level of chemical substances and the intensity of earthquakes. They are also useful as inverses of the exponential functions. All this is discussed in the sections that follow.

Exponential Functions

An exponential function is one that has the general form $f(x) = a^x$. The variable, x, is the exponent or power, and the base, a, is a constant. The constant must always be a positive number, but not equal to 1. All of the rules of exponents apply to the exponential functions.

$$a^x \cdot a^y = a^{x+y} \qquad \frac{a^x}{a^y} = a^{x-y} \qquad \left(a^x\right)^y = a^{x \cdot y} \qquad a^{-x} = \frac{1}{a^x}$$

Look at the graphs of some exponential functions. The generalizations about them follow.

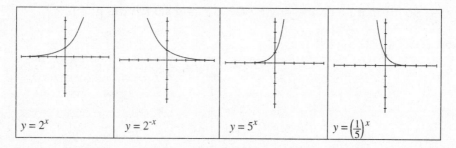

$y = 2^x \qquad\qquad y = 2^{-x} \qquad\qquad y = 5^x \qquad\qquad y = \left(\frac{1}{5}\right)^x$

The graph of an exponential function either continuously rises or continuously falls. When does it rise? When does it fall? Why do some exponential functions rise or fall more rapidly than others? The things to consider are whether the constant base a is between 0 and 1 or greater than 1. Another consideration is whether x has a positive or negative sign.

Form $y = a^x$

If the exponential equation is $y = a^x$, and $a > 1$, then the graph of the function rises as you move from left to right. The greater the value of a, the more quickly it rises. Compare the graphs of $y = 2^x$ and $y = 5^x$, in the preceding figure. They both cross at the point $(0,1)$, because $2^0 = 1$, $5^0 = 1$, and, in general, $a^0 = 1$. All exponential functions of this form have a y-intercept at $(0,1)$.

If the exponential equation is $y = a^x$, and $0 < a < 1$, which means that a can be some proper fraction, then the graph falls as you move from left to right. Look at the graph of $y = \left(\frac{1}{5}\right)^x$ in the figure. The greater the value of the denominator in the fraction (the closer the fraction is to 0), the more steeply the graph falls.

In both cases, the graph of the function seems to hug the x-axis, either to the left or to the right. The values of the function get very close to 0, but they never actually become 0.

Form $y = a^{-x}$

If the exponential equation is $y = a^{-x}$, and $a > 1$, then the graph of the function falls as you move from left to right. The greater the value of a, the more quickly it falls. Look at the graph of $y = 2^{-x}$. It crosses at the point $(0,1)$ and hugs the x-axis as you move to the right.

If the exponential equation is $y = a^{-x}$, and $0 < a < 1$, then the graph rises as you move from left to right. This type of function has a double negative at work. Consider, for instance, the function $y = \left(\frac{1}{3}\right)^{-x}$. Using the laws of exponents to change the fraction, $\left(\frac{1}{3}\right) = 3^{-1}$. Substitute that in the original equation, $y = \left(\frac{1}{3}\right)^{-x} = \left(3^{-1}\right)^{-x} = 3^{-1(-x)} = 3^x$. You're back to a function with a positive exponent and a base greater than 1.

Example Problems

These problems show the answers and solutions.

1. Find three points that lie on the graph of $y = 3^x$ and sketch the graph.

    answer: $(0,1)$, $(2,9)$, $\left(-1,\frac{1}{3}\right)$

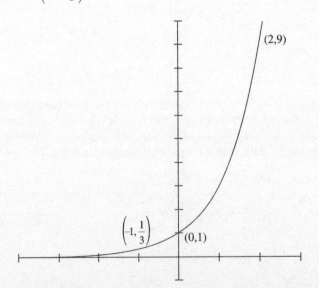

To find points that satisfy the function equation, make a chart of x values.

x	y
0	$3^0 = 1$
2	$3^2 = 9$
−1	$3^{-1} = \dfrac{1}{3}$

2. Find three points that lie on the graph of $y = \left(\dfrac{1}{4}\right)^x$ and sketch the graph.

answer: $(0, 1)$, $\left(2, \dfrac{1}{16}\right)$, $(-1, 4)$

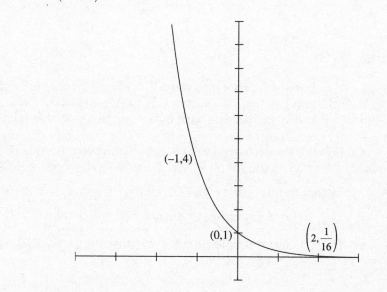

To find points that satisfy the function equation, make a chart of x values.

x	y
0	$\left(\dfrac{1}{4}\right)^0 = 1$
2	$\left(\dfrac{1}{4}\right)^2 = \dfrac{1^2}{4^2} = \dfrac{1}{16}$
−1	$\left(\dfrac{1}{4}\right)^{-1} = \left(4^{-1}\right)^{-1} = 4$

Work Problems

Use these problems to give yourself additional practice.

Match each equation with its corresponding graph.

1. $y = 2.3^x$

2. $y = \left(\dfrac{1}{2}\right)^x$

3. $y = 8^{-x}$

4. $y = \left(\dfrac{2}{3}\right)^{x}$

5. $y = \left(\dfrac{1}{5}\right)^{-x}$

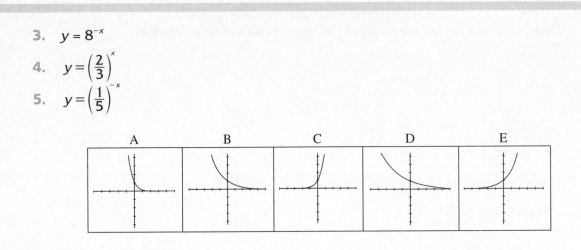

Worked Solutions

1. **E**   Graphs C and E rise as they go from left to right. The choice here is E, because it doesn't rise as quickly (steeply) as graph C. Graph C rises more quickly, because its corresponding equation has a larger base; a base of 5 is greater than this base of 2.3.

2. **B**   Graphs A, B, and D fall as they go from left to right. The choice here is B, because it doesn't fall as quickly (steeply) as the graph of $y = 8^{-x}$, and it falls more quickly than the graph of $y = \left(\dfrac{2}{3}\right)^{x}$. To compare these three graphs, change them all to the form $y = a^{-x}$. The greater the value of a, the more rapidly (steeper) it will fall: $y = \left(\dfrac{1}{2}\right)^{x} = 2^{-x}$, $y = \left(\dfrac{2}{3}\right)^{x} = \left(\dfrac{3}{2}\right)^{-x}$, $y = 8^{-x}$. So, in order of least steep to most steep, you have $\left(\dfrac{3}{2}\right)^{-x}$, 2^{-x}, and 8^{-x}.

3. **A**   A is the steepest, going down from left to right. Refer to the explanation for problem 2.

4. **D**   D is the least steep, going down from left to right. Refer to the explanation for problem 2.

5. **C**   C is the steepest, going up from left to right. It can be rewritten $y = \left(\dfrac{1}{5}\right)^{-x} = \left(5^{-1}\right)^{-x} = 5^{x}$.

The Constant "e"

The lowercase letter *e* represents a constant value that is used extensively in science, engineering, social sciences, and, of course, mathematics. The approximate value of *e* is 2.718. If you want more decimal places than that, the first 22 decimal places are 2.7182818284590452353603. The value of *e* has been calculated to millions of decimal places, but that's much more than necessary here.

e is used so much because it occurs naturally in many formulas and relationships. It's found in formulas for exponential growth and decay, in the statistical bell-shaped curve, in the shape of a hanging cable, in probability, and many more. *e* is like π, the relationship between the circumference and diameter of a circle, in that it has a decimal that goes on forever and ever without repeating or terminating. Both of these constants are important for doing computations.

Decimal approximations of e can be obtained in one of several ways. One way is to compute the value of $\left(1 + \frac{1}{n}\right)^n$, letting n get larger and larger. The greater the value of n, the more decimal places of e can be determined. Of course, it takes a computer to be able to determine those values. Another way of computing e is to add up the sum $\frac{1}{0!} + \frac{1}{1!} + \frac{1}{2!} + \frac{1}{3!} + \ldots + \frac{1}{n!}$ letting n get larger and larger. The greater the value of n, the more decimal places.

When e occurs as the base of an exponential function, $y = e^x$, it follows the same rules as any other number that lies between 2 and 3. The graph of $y = e^x$ rises as you read from left to right, and the graph of $y = e^{-x}$ falls. The graph of $y = e^x$ is steeper than the graph of $y = 2^x$ and not as steep as the graph of $y = 3^x$. It crosses the y-axis at the point $(0,1)$. Exponential expressions involving e follow the same rules as other bases.

$$e^x \cdot e^y = e^{x+y} \qquad \frac{e^x}{e^y} = e^{x-y} \qquad \left(e^x\right)^y = e^{x \cdot y} \qquad e^{-x} = \frac{1}{e^x}$$

Example Problems

These problems show the answers and solutions.

1. Simplify $\left(e^x\right)^2$.

    answer: e^{2x}

    When raising a power to a power, multiply the two powers together.

2. Simplify $e^x \cdot e^2$.

    answer: e^{x+2}

    When multiplying two values with the same base, add their exponents.

3. Simplify $\left(\frac{1}{e^x}\right)^{-2}$.

    answer: e^{2x}

    First write the fraction using a negative exponent. Then follow the rule for raising a power to a power: $\left(\frac{1}{e^x}\right)^{-2} = \left(e^{-x}\right) = e^{-x(-2)} = e^{2x}$.

4. Put the following in order from smallest to largest value: $3^{100}, \left(\frac{1}{3}\right)^{100}, e^{100}, e^{-100}$.

    answer: $\left(\frac{1}{3}\right)^{100}, e^{-100}, e^{100}, 3^{100}$

    The easiest way to compare them is to put them all in the same form. In this case, change all the terms to the form a^x, so that all the exponents are positive: $3^{100}, \left(\frac{1}{3}\right)^{100}, e^{100}, \left(\frac{1}{e}\right)^{100}$.

    The only term that changed was the last term. Now replace the e's with the approximation 2.718: $3^{100}, \left(\frac{1}{3}\right)^{100}, 2.718^{100}, \left(\frac{1}{2.718}\right)^{100}$. They're all raised to the same power, so the smallest will be the fraction with the largest denominator, and the largest will be the number that isn't a fraction and has the largest base. In order from smallest to largest, you have $\left(\frac{1}{3}\right)^{100}$, $\left(\frac{1}{2.718}\right)^{100}$, 2.718^{100}, 3^{100}. Change the 2.718s back to e's to get close to the original form.

Work Problems

Use these problems to give yourself additional practice.

1. Simplify $\dfrac{e^4}{e^x}$.

2. Simplify $\left(e^{2x}\right)^{-3x}$.

3. Simplify $e^3\left(e^{x+1}\right)$.

4. Simplify $e^x\left(e^{2x}\right)^2$.

5. Arrange from the smallest to the largest: e^3, 2^3, $\left(\dfrac{1}{e}\right)^{-2.9}$.

Worked Solutions

1. e^{4-x} When two terms with the same base are divided, you subtract the exponents.

2. $\dfrac{1}{e^{6x^2}}$ First multiply the two powers to get e^{-6x^2}. Then rewrite the expression with a positive exponent by moving the term to the denominator.

3. e^{x+4} Add exponents.

4. e^{5x} First raise the power to a power; then add the exponents: $e^x\left(e^{2x}\right)^2 = e^x\left(e^{4x}\right) = e^{5x}$.

5. 2^3, $\left(\dfrac{1}{e}\right)^{-2.9}$, e^3 Rewrite these with positive exponents and replace the e's with 2.718.

   2^3, $(e^{-1})^{-2.9} = 2.718^{2.9}$, 2.718^3. 2^3 is equal to 8. 2.718 is close to 3, so raising that to almost the third power will result in a number much bigger than 8.

Compound Interest

An application of exponential functions that everyone can relate to is that of compound interest. When you deposit your money in an account for an investment, you expect the amount in the account to grow over time—if you don't remove any of it. A formula for determining the total amount of money in an account is $A = P\left(1 + \frac{r}{n}\right)^{nt}$. The P stands for the principal, or the amount of money that is deposited in the beginning. The r stands for the interest rate, and it is entered as a decimal value. The n stands for the number of times each year that the amount in the account is compounded. And the t stands for the number of years that the money stays in the account.

The result A could be a function of the rate, of the number of compounding times, of the time, and of the principal. Any and all of these can be variables. You'll usually see this formula written as a function of time, though. For instance, if you invest $10,000 at $4\frac{1}{4}$ percent compounded quarterly (4 times each year), then the function reads $A = 10,000\left(1 + \frac{.0425}{4}\right)^{4t}$ where the only input variable is t, the amount of time. This formula can be simplified, using rules for simplifying exponents, $A = 10,000\left(1 + \frac{.0425}{4}\right)^{4t} = 10,000\left[(1 + .010625)^4\right]^t = 10,000[1.043182]^t$. Now the base of the exponential function is 1.043182; the exponent is t, the number of years; and the 10,000 is a constant multiplier of the result.

What if another investment institution says that you can make more money with them, because they will raise the interest rate to $4\frac{3}{8}$ percent? The only other change is that it's compounded only twice each year instead of four times each year. Will there be much difference? Is this really better? Go back to the original formula and plug in the numbers. $A = 10,000\left(1 + \frac{0.4375}{2}\right)^{2t}$ looks better with the higher interest rate, but there's still the question of the compounding periods. Simplify this, $A = 10,000\left(1 + \frac{.04375}{2}\right)^{2t} = 10,000\left[(1.021875)^2\right]^t = 10,000[1.044229]^t$. The base is larger. This appears to be a better situation. Now compare the actual amounts in the account after 10 years to see what kind of difference this makes.

$$\text{Using } A = 10,000[1.043182]^t, A = 10,000[1.043182]^{10} = 15,261.63.$$
$$\text{Using } A = 10,000[1.044229]^t, A = 10,000[1.044229]^{10} = 15,415.50.$$

There's a difference of about $154 over the 10-year period. That could be considered to be significant, but, then again, if the first institution has a better location and other services, it might be better to stay with it.

These situations involve exponential functions that compound twice a year or four times a year or twelve times a year, and so on. These are relatively small numbers of times compared to compounding *continuously*. Compounding continuously means that the compounding is occurring infinitely many times. That's sort of hard to imagine. Look at what this does to the compound interest formula.

Starting with $A = P\left(1 + \frac{r}{n}\right)^{nt}$, rewrite it as $A = P\left[\left(1 + \frac{r}{n}\right)^n\right]^t$. In the earlier section on "The Constant 'e'", one of the ways of computing e is to use $\left(1 + \frac{1}{n}\right)^n$ and let n get infinitely large. Then $\left(1 + \frac{1}{n}\right)^n \to e$. Do the same thing with $\left(1 + \frac{r}{n}\right)^n$, and the r makes $\left(1 + \frac{r}{n}\right)^n \to e^r$. Replace that in the formula, and $A = P\left(e^r\right)^t = Pe^{rt}$. This is the formula for continuous compounding.

Example Problems

These problems show the answers and solutions.

1. Compute the total amount of an investment after four years if the original amount deposited was $5,000, the interest rate was 2 percent, and it compounded monthly.

    answer: $5,416.07

    Use the formula $A = P\left(1 + \frac{r}{n}\right)^{nt}$. $A = 5,000\left(1 + \frac{.02}{12}\right)^{12 \cdot 4}$.

2. Compute the total amount of an investment after 15 years if the original amount deposited was $100,000, the interest rate was $3\frac{1}{2}$ percent, and it compounded continuously.

   answer: $169,045.88

   Use the formula $A = Pe^{rt}$. $A = 10,000e^{.035 \cdot 15}$.

Work Problems

Use these problems to give yourself additional practice.

1. Determine the total amount of the investment if $800 is invested at $4\frac{3}{4}$ percent for 5 years compounded annually.

2. Determine the total amount of the investment if $80,000 is invested at $2\frac{5}{8}$ percent for 6 months compounded quarterly.

3. Determine the total amount of the investment if $50 is invested at 6 percent for 20 years compounded continuously.

4. Determine the total amount of the investment if $15,000 is invested at 0.9 percent for 6 years compounded continuously.

5. In 2002, you received a letter from the First Bank of the West Indies informing you that one of your ancestors stopped there with Columbus in 1492 and deposited $1. Unfortunately, this ancestor sank with his ship and never returned. That $1 has been earning interest at the rate of 3 percent, compounded semiannually. It's been an inactive account for a long time, now, and they'd like you to either claim it or sign off on it. What is the account worth? Should you claim it?

Worked Solutions

1. **$1008.93**   Use the formula $A = P\left(1 + \frac{r}{n}\right)^{nt} = 800\left(1 + \frac{.0475}{1}\right)^{1 \cdot 5}$. Compounded *annually* means that it's compounded just once every year.

2. **$81,053.45**   Use the formula $A = P\left(1 + \frac{r}{n}\right)^{nt} = 80,000\left(1 + \frac{.02625}{4}\right)^{4 \cdot 0.5}$. Compounded quarterly means that it's compounded four times each year. The time is 6 months, which is only half or 0.5 of one year.

3. **$166.01**   Use the formula $A = Pe^{rt} = 50e^{0.06 \cdot 20}$.

4. **$15,832.27**   Use the formula $A = Pe^{rt} = 15,000e^{0.09 \cdot 6}$. The percentage 0.9 percent means it's less than 1 percent. You move the decimal place two places to the left.

5. **$3,938,792.43**   I certainly hope you planned to keep it! Use the formula $A = P\left(1 + \frac{r}{n}\right)^{nt} = 1\left(1 + \frac{.03}{2}\right)^{2 \cdot 511}$. Ah, the power of compounding!

Logarithmic Functions

Logarithmic functions have their place as models of some real-life situations, but their biggest claim to fame is as inverses of exponential functions.

Functions that have inverses must be one-to-one. Refer to "Inverse Functions" in Chapter 3 for more on this. Exponential functions are one-to-one and so have inverses. Use the method used to solve for the inverse of a function, found in "Inverse Functions" to find the inverse of the exponential function $f(x) = 3^x$. First rewrite it as $y = 3^x$ and then exchange the x's and the y's, giving you $x = 3^y$. Now solve for y. But how can you do that? The y is in the exponent, so you can't add, subtract, multiply, divide, or try to take a root of each side. You could just solve for y by saying, "$y =$ the exponent that you raise the 3 to in order to get x." Replace "the exponent that you raise the 3 to" with $\log_3$, and put the x after it so you'll have $y = \log_3 x$. The word "log" is short for logarithm. So the inverse of $f(x) = 3^x$ is $f^{-1}(x) = \log_3 x$. The inverse of $g(x) = 8^x$ is $g^{-1}(x) = \log_8 x$, and so on. An exponential function, such as $f(x) = 3^x$ has a domain of all real numbers (x can be anything), and a range of $y > 0$. The y values are never negative or 0. The inverse of an exponential function has these two things reversed. The domain of a log function is $x > 0$. The input has to be numbers bigger than 0. And, the range is all real numbers.

To solve for the output values of logarithmic functions, two equivalences are very nice to know:

1. $\log_a x = y \Leftrightarrow a^y = x$

2.   If $a^x = a^y$, then $x = y$.

Example Problems

These problems show the answers and solutions.

1. Solve for the value of $\log_2 8$.

    answer: 3

    The exponent that 2 is raised to in order to get an 8 is 3. Change from $\log_2 8 = x$ to $2^x = 8$. Then rewrite the 8 as a power of 2. $2^x = 2^3$. So $x = 3$.

2. Solve for the value of $\log_3 \frac{1}{3}$.

    answer: -1

    The exponent that 3 is raised to in order to get $\frac{1}{3}$ is -1. Change from $\log_3 \frac{1}{3} = x$ to $3^x = \frac{1}{3}$. Then rewrite the $\frac{1}{3}$ as a power of 3. $3^x = \frac{1}{3} = 3^{-1}$. So $x = -1$.

3. Solve for the value of $\log_9 \sqrt{3}$.

    answer: $\frac{1}{4}$

    Change from $\log_9 \sqrt{3} = x$ to $9^x = \sqrt{3}$. Then write each term as a power of 3: $\left(3^2\right)^x = 3^{1/2}$, $3^{2x} = 3^{1/2}$. Now set the exponents equal to one another and solve for x: $2x = \frac{1}{2}$, $x = \frac{1}{4}$.

    A very important logarithm is $\log_e x$, which is abbreviated $\ln x$ and is called the *natural logarithm*. This is an important logarithm, because powers of e and log base e are used in so many formulas and relationships. Solving logarithm expressions with e in them is a bit harder, because powers of 2.71828 aren't very nice to deal with.

Scientific calculators have a log function and an ln function. These are used to evaluate problems such as log 4 or ln 3. But what base is log, with no subscript? It's assumed that you're dealing with $\log_{10}$. Logarithms with base 10 are called *common logarithms*. They were developed first, because we have a base 10 number system, and we deal in powers of 10. Computers and calculators have made other computations in other bases possible, but we still go back to log base 10. If you want to use a calculator to find a value that isn't in $\log_{10} = \log$ or $\log_e = \ln$, then you can use a formula to do the computation. The formula is $\log_b x = \dfrac{\log x}{\log b}$ or $\dfrac{\ln x}{\ln b}$. You'll still need a calculator to do the computation, but at least it's possible to find an answer.

Example Problems

These problems show the answers and solutions.

1. Use a scientific calculator to find log 40.

   answer: 1.60205999 (to eight places)

   This tells you that if you raise 10 to the power 1.60205999, you'll get 40.

2. Use a scientific calculator to find ln 6.

   answer: 1.79175947

   This tells you that if you raise e to the power 1.79175947, you'll get 6.

3. Use a scientific calculator to find $\log_2 8$.

   answer: 3

   Using a calculator and the formula, $\log_2 8 = \dfrac{\ln 8}{\ln 2}$, you'll get 3. Of course, you could have done this one without a calculator, because 8 is a power of 2. Just rewrite $\log_2 8 = x$ as $2^x = 8 = 2^3$.

Work Problems

Use these problems to give yourself additional practice.

1. Solve $\log_4 64$.

2. Solve $\log_8 32$.

3. Solve $\log_9 \dfrac{1}{27}$.

4. Solve ln e.

5. Solve ln 1.

Worked Solutions

1. **3**   Change $\log_4 64 = x$ to $4^x = 64 = 4^3$. Set the exponents equal; $x = 3$.

2. $\frac{5}{3}$ Change $\log_8 32 = x$ to $8^x = 32$. Then rewrite each number as a power of 2: $(2^x)^3 = 2^5$. $2^{3x} = 2^5$. When the two exponents are set equal to one another, $3x = 5$. Then just solve for x.

3. $-\frac{3}{2}$ Change $\log_9 \frac{1}{27} = x$ to $9^x = \frac{1}{27}$. Then rewrite each number as a power of 3, giving you $(3^2)^x = \frac{1}{3^3} = 3^{-3}$; $3^{2x} = 3^{-3}$. Setting the exponents equal to one another, $2x = -3$, and solve for x.

4. **1**   Change $\ln e = x$ to $\log_e e = x$ and then rewrite it as $e^x = e$. The value of x must be 1.

5. **0**   Change $\ln 1 = x$ to $\log_e 1 = x$ and then rewrite it as $e^x = 1$. The value of x must be 0, because $e^0 = 1$.

Laws of Logarithms

Logarithms are functions that are closely tied to exponential functions. There are rules for simplifying and manipulating logarithmic expressions that somewhat resemble the rules involving exponents. For instance, you can simplify $\log_2 8$. This isn't a hard problem just by itself. Since $2^3 = 8$, the answer to this log problem is $\log_2 8 = 3$. If you need more information on how to do this, go back to the previous section. Another way of writing $\log_2 8$ is $\log_2 4 \cdot 2 = \log_2 4 + \log_2 2$. Because you know that $\log_2 4 = 2$ and $\log_2 2 = 1$, then $\log_2 4 + \log_2 2 = 2 + 1 = 3$. Obviously, such a simplification is not necessary here. It's reserved for when it's really needed. Here are some rules for simplifying or changing logarithmic expressions. The examples will provide some situations in which the rules seem to be more necessary.

Laws

1. $\log_a(x \cdot y) = \log_a x + \log_a y$ The log of a product is equal to the sum of the logs.

2. $\log_a \frac{x}{y} = \log_a x - \log_a y$ The log of a quotient is equal to the difference between the logs.

3. $\log_a(x^n) = n \cdot \log_a x$ The log of a power is equal to the product of that power and the log.

4. $\log_a a = 1$ When the base and the number being evaluated are the same, the result is 1.

5. $\log_a 1 = 0$ When the number being evaluated is 1, the result is 0.

6. $\log_a \frac{1}{x} = -\log_a x$ The log of the reciprocal of x is negative log x.

Simplifying expressions with logarithms in them can result in terms that are much simpler. In many cases, you can rewrite the original expression without powers and products and quotients. This is helpful when you need to combine different expressions. It's even more helpful in calculus problems in which a derivative is made much simpler when the logarithm is rewritten.

The Laws of Logarithms are very helpful, but you can make some common errors if you aren't careful. For instance,

$$\log_a(x + y) \neq \log_a x + \log_a y \quad \text{Just look at Law 1.}$$
$$\log_a(x - y) \neq \log_a x - \log_a y \quad \text{Look at Law 2.}$$

$$\log_a x - \log_a y - \log_a z = \log_a \frac{\left(\frac{x}{y}\right)}{z} \; BUT \; \log_a x - \log_a y - \log_a z \neq \log_a \frac{x}{\left(\frac{y}{z}\right)}$$

$$\log_a x - \left(\log_a y + \log_a z\right)$$

Another way to have simplified $\log_a x - \log_a y - \log_a z$ is to rewrite it as $= \log_a x - \log_a(yz)$

$$= \log_a \frac{x}{yz}$$

Example Problems

These problems show the answers and solutions.

1. Use laws of logarithms to simplify $\log_2(4x^2)$.

    answer: $2 + 2\log_2 x$

    First, write the expression as the sum of the logs of the two factors: $\log_2 4x^2 = \log_2 4 + \log_2 x^2$. Write the 4 in the first term as 2^2. Then simplify both terms using the law for the log of a power. $\log_2 2^2 + \log_2 x^2 = 2\log_2 2 + 2\log_2 x$. Using the law involving the base and number evaluated being the same, the value of $2\log_2 2 = 2(1) = 2$. Replace the first term with the 2, and the answer is complete.

2. Use laws of logarithms to simplify $\log_6 x^3 \sqrt{x + 1}$.

    answer: $3\log_6 x + \frac{1}{2}\log_6(x + 1)$

    First, write the expression as the sum of the logs of the two factors. Change the radical to a $\frac{1}{2}$ power: $\log_6 x^3 \sqrt{x + 1} = \log_6 x^3 + \log_6(x + 1)^{1/2}$. Now use the law involving the log of a power: $\log_6 x^3 + \log_6(x + 1)^{1/2} = 3\log_6 x + \frac{1}{2}\log_6(x + 1)$.

3. Use laws of logarithms to simplify $\log_3 \frac{3x}{(x - 7)^4}$.

    answer: $1 + \log_3 x - 4\log_3(x - 7)$

    First, write the expression as the difference of the logs of the numerator and denominator: $\log_3 \frac{3x}{(x - 7)^4} = \log_3 3x - \log_3(x - 7)^4$. The first term has a product, so it can be written as the sum of the logs of the factors. The second term has a power, so it can be rewritten, giving you $\log_3 3 + \log_3 x - 4\log_3(x - 7)$. The last thing to do is to replace the first term with 1.

4. Use laws of logarithms to simplify $\ln \frac{1}{5x^2 e^x}$.

    answer: $-\ln 5 - 2\ln x - x$

    First, use the law involving a reciprocal to rewrite the expression as $-\ln 5x^2 e^x$. Rewrite the product as the sum of the factors, leaving the negative sign outside a parenthesis: $-\ln 5x^2 e^x = -(\ln 5 + \ln x^2 + \ln e^x)$. The second and third terms can be simplified using the law involving powers: $-(\ln 5 + 2\ln x + x \ln e)$. The value of $\ln e$ is 1. This uses the law where the base of the logarithm and the expression being evaluated are the same. Replace the $\ln e$ with 1 and distribute the negative sign: $-(\ln 5 + 2 \ln x + x \ln e) = -\ln 5 - x \ln x - x$.

Work Problems

Use these problems to give yourself additional practice.

1. Use laws of logarithms to simplify $\log_2 16x^2$.

2. Use laws of logarithms to simplify $\log_a \dfrac{(x^2-4)^3}{\sqrt{x}}$.

3. Use laws of logarithms to simplify $\log \dfrac{500}{x}$.

4. Use laws of logarithms to simplify $\ln x(6+x)^8$.

5. Use laws of logarithms to simplify $\ln 4xe^{2x}$.

Worked Solutions

1. **$4 + 2\log_2 x$**   First use the law for the log of a product. Then rewrite the 16 as a power of 2: $\log_2 16x^2 = \log_2 16 + \log_2 x^2 = \log_2 2^4 + \log_2 x^2$. Now apply the law for the log of a power to both terms: $4\log_2 2 + 2\log_2 x$. The value of $\log_2 2 = 1$, so replace that in the expression to get $4(1) + 2\log_2 x = 4 + 2\log_2 x$.

2. **$3\log_a(x+2) + 3\log_a(x-2) - \dfrac{1}{2}\log_a x$**   First, use the law for the log of a quotient to get $\log_a \dfrac{(x^2-4)^3}{\sqrt{x}} = \log_a(x^2-4)^3 - \log_a \sqrt{x} = \log_a(x^2-4)^3 - \log_a x^{1/2}$. Use the law for the log of a power to rewrite both terms: $3\log_a(x^2-4) - \dfrac{1}{2}\log_a x$. Now factor the $x^2 - 4$ in the first term and apply the law involving products: $3\log_a(x+2)(x-2) - \dfrac{1}{2}\log_a x = 3\left[\log_a(x+2) + \log_a(x-2)\right] - \dfrac{1}{2}\log_a x$. Now distribute the 3 to get the final answer.

3. **$\log 5 + 2 - \log x$**   First apply the law involving the log of a quotient to get $\log \dfrac{500}{x} = \log 500 - \log x$. Now rewrite the 500 as the product of 5 and 100 and apply the law involving a product: $\log 500 - \log x = \log 5 \cdot 100 - \log x = \log 5 + \log 100 - \log x$. The number 100 is equal to 10^2, so replace the 100 and apply the law for the log of a power: $\log 5 + \log 10^2 - \log x = \log 5 + 2\log 10 - \log x$. When there is no base shown, it's assumed that the log is base 10, so the term $2\log 10$ is actually $2\log_{10} 10$. Apply the law for the base and number evaluated being the same to replace that factor with 1: $\log 5 + 2\log 10 - \log x = \log 5 + 2(1) - \log x$.

4. **$\ln x + 8\ln(6+x)$**   Apply the law for the log of a product to get $\ln x(6+x)^8 = \ln x + \ln(6+x)^8$. Now apply the law involving the log of a power to rewrite the second term.

5. **$\ln 4 + \ln x + 2x$**   Apply the law for the log of a product to get $\ln 4xe^{2x} = \ln 4 + \ln x + \ln e^{2x}$. The last term involves a product and can be written $\ln e^{2x} = 2x \ln e$. The value of $\ln e$ is 1. Replace that in the expression to finish the problem.

Applications of Logarithms

Logarithmic functions are found as models in many scientific and engineering applications. Base 10 and base e logs are used, and a scientific calculator becomes very important when working with these logs. Some of the applications are shown here as example problems.

Example Problems
These problems show the answers and solutions.

1. The intensity of an earthquake is measured using the Richter Scale. This is based on the logarithmic function, $I = \log \frac{A}{P}$. The I is the intensity of the earthquake—it's the number you hear reported when an earthquake occurs. The A is the amplitude measured in micrometers, and the P is the period measured in seconds. If the amplitude of an earthquake is 6,000 micrometers, and the period is 0.08 second, then what is the intensity?

    answer: 4.9

    This answer is rounded to the nearer tenth. Log base 10 is assumed, because there's no other base shown. Using a calculator, $\log \frac{6,000}{0.08} = 4.875061263$.

2. Refer to the first example. What would be the intensity of an aftershock, if the period was the same but the amplitude was *half* that of the original earthquake's?

    answer: 4.6

    I'll bet you thought it would be a lot less! Here is the computation:
    $\log \frac{3,000}{0.08} = 4.574031268$. The answer is rounded to one decimal place.

3. The loudness of sound is measured in decibels and is defined as $L = 10 \log \frac{I}{I_0}$. The L is the number of decibels. The I is the intensity of the sound measured in watts per square meter, and I_0 is equal to 10^{-12} watts per square meter. This is the least intense sound that a human ear can detect. A loudness of 140 decibels, which is about that of a shotgun blast, is painful to the human ear. A measure of 40 decibels, which is about what a running refrigerator will produce, is a relatively low measure. If a firecracker is measured at 120 decibels, and a vacuum is measured at 80 decibels, how much more intense is the sound from the firecracker?

    answer: 10,000 times as intense

    The firecracker is 120 decibels: $120 = 10 \log \frac{f}{I_0}$ where f is the intensity of that sound. The vacuum cleaner is 80 decibels: $80 = 10 \log \frac{v}{I_0}$ where v is the intensity of that sound. Solving for the intensity of each, first, when $120 = 10 \log \frac{f}{I_0}$ divide each side by 10: $12 = \log \frac{f}{I_0}$. Now rewrite this as an exponential expression in base 10: $10^{12} = \frac{f}{I_0}$. Then multiply each side by I_0. $f = 10^{12} I_0$. Do the same with the other, and $v = 10^8 I_0$. f is much larger than v. In fact, to get $10^{12} I_0$ from $10^8 I_0$, you have to multiply by 10^4. That's 10,000, or f is 10,000 times as much as v.

4. The *doubling time* for something that's growing exponentially is given by the formula $t = \frac{\ln 2}{r}$. The t is the time, and the r is the growth rate for that time. If the population in a city is growing exponentially at the rate of 5 percent per year, then how long will it take for the population to double?

    answer: about 14 years

    Using the formula, $t = \frac{\ln 2}{0.05} = 13.8629$.

5. If a bacterial culture is growing exponentially and the number of bacteria doubles in 6 hours, then what is the hourly growth rate?

    answer: about 11.6 percent

    Starting with the formula, $t = \frac{\ln 2}{r}$, multiply each side by r and divide each side by t to get $r = \frac{\ln 2}{t}$. Now $r = \frac{\ln 2}{6} = .115525$.

6. When recharging a battery, the rate at which it charges depends on how close the battery is to being fully charged. If the charge is very low, then it recharges more rapidly than when it's close to being fully charged. The formula used to determine the amount of time needed to charge a battery to a certain level is $t = -\frac{1}{k} \ln\left(1 - \frac{C}{M}\right)$. The t is the number of minutes. The k is a positive constant that is special to this particular battery and this particular charger. The C is the level that you want the battery charged to, and the M is the maximum charge that the battery can hold. How long will it take to bring a fully discharged battery to 95 percent of full charge if $k = 0.03$?

    answer: 100 minutes

    Using the formula, $t = -\frac{1}{0.03} \ln\left(1 - \frac{.95M}{M}\right)$. If you want 95 percent of the possible full charge, that's $.95M$, which replaced the C. So the M's divide through leaving just the 0.95 to be subtracted, $t = -\frac{1}{0.03} \ln\left(1 - \frac{.95M}{M}\right) = -\frac{1}{0.03} \ln(1 - .95) = -\frac{1}{0.03} \ln(0.05) = 99.8577$ or about 100 minutes.

7. Referring to the problem given previously, how long would it take if you only wanted a 90 percent charge?

    answer: about 77 minutes

    The formula would be $t = -\frac{1}{0.03} \ln\left(1 - \frac{.90M}{M}\right) = -\frac{1}{0.03} \ln(1 - .90)$ $= -\frac{1}{0.03} \ln(0.1) = 76.7528$.

Work Problems

Use these problems to give yourself additional practice.

1. What is the intensity of an earthquake with an amplitude of 7,000 micrometers and a period of 0.05 seconds?

2. What is the intensity of an aftershock of the earthquake described in problem 1 if the amplitude is half that of the original?

3. How much more intense is music at 110 decibels than normal conversation that's at 50 decibels?

4. How long will it take the population of a city to double if it's growing exponentially at the rate of $2\frac{1}{2}$ percent per year?

5. How long will it take for a battery to charge to 90 percent of full charge if $k = 0.02$?

Worked Solutions

1. **5.1**   Using the formula, $I = \log \frac{A}{P} = \log \frac{7,000}{0.05} = 5.146$.

2. **4.8**   Using the formula, $I = \log \frac{A}{P} = \log \frac{3,500}{0.05} = 4.845$.

3. **1,000,000**   The music at 110 decibels has intensity m, where $110 = 10\log \frac{m}{I_0}$. Dividing by 10 and rewriting this as an exponential expression, $10^{11} = \frac{m}{I_0}$ or $m = 10^{11}I_0$. Doing the same thing for the conversation, its intensity is c: $50 = 10\log \frac{c}{I_0}$. Dividing by 10 and rewriting this as an exponential expression, $10^5 = \frac{c}{I_0}$ or $c = 10^5 I_0$. You'll have to multiply by 10^6 or 1,000,000 to raise the intensity of the conversation to that of the music.

4. **about 28 years**   The formula is $t = \frac{\ln 2}{r} = \frac{\ln 2}{0.025} = 27.72589$.

5. **about 115 minutes**   The formula is $t = -\frac{1}{k}\ln\left(1 - \frac{C}{M}\right) = -\frac{1}{0.02}\ln\left(1 - \frac{0.9M}{M}\right) = -\frac{1}{0.02}\ln(1 - 0.9) = 115.129$.

Solving Exponential and Logarithmic Equations

Many situations occur in which solving a problem or application turns into solving an equation involving exponential functions or logarithmic functions or both. The laws for simplifying exponential expressions (found in the section on "Exponential Functions" in this chapter) and the laws for simplifying logarithmic expressions (found in the section on "Laws of Logarithms" in this chapter) may be called into play to complete these problems. Three properties that are used frequently to do these problems are equivalences between logarithms and exponentials and algebraic expressions.

1. $\log_a x = y \Leftrightarrow a^y = x$
2. $a^x = a^y \Leftrightarrow x = y$ (assume a is positive)
3. $\log_a x = \log_a y \Leftrightarrow x = y$

Example Problems

1.   Solve for x in $4^{3-x} = 2$.

   answer: $\frac{5}{2}$

   First rewrite the 4 as a power of 2 and multiply the two powers together. The power of the 2 on the right is implied to be a 1: $\left(2^2\right)^{3-x} = 2$, $2^{6-2x} = 2^1$. Now set the two exponents of the 2s equal to one another: $6 - 2x = 1$. Solve for x. $-2x = -5$, $x = \frac{5}{2}$.

2.   Solve for x in $27^{x^2+x} = 3^{x+8}$.

   answer: $-2, \frac{4}{3}$

First rewrite the 27 as a power of 3 and multiply the exponents: $\left(3^3\right)^{x^2+x} = 3^{x+8}$, $3^{3x^2+3x} = 3^{x+8}$. Now set the two exponents equal to one another to get $3x^2 + 3x = x + 8$. Set this equal to 0 by moving everything over to the left, and you get a quadratic equation that can be factored and solved: $3x^2 + 3x - x - 8 = 3x^2 + 2x - 8 = (3x - 4)(x + 2) = 0$. When $3x - 4 = 0$, $x = \frac{4}{3}$. When $x + 2 = 0$, $x = -2$.

3. Solve for y in $\ln y^2 - \ln 4 = \ln 2y$.

   answer: 8

   First, apply the law of logarithms involving the log of a quotient. This is going backward, changing from the difference of two logs to the quotient: $\ln \frac{y^2}{4} = \ln 2y$. Now the two expressions in the logarithms can be set equal to one another: $\frac{y^2}{4} = 2y$. Multiply each side by 4 and then set the quadratic equal to 0 by moving everything over to the left: $y^2 = 8y$, $y^2 - 8y = 0$. This factors into $y(y - 8) = 0$. The two factors, set equal to 0, give you $y = 0$ and $y = 8$. The $y = 0$ cannot be used. It's called an extraneous root, because it is a solution of the quadratic, but it is not a solution of the original log problem. The reason it doesn't work is because, if you put it back into the original problem, you get $\ln 0$. Logarithms have domains (input values) of positive numbers only—no 0s or negatives.

4. Solve for x in $\log_5 \sqrt{x - 4} = 2$.

   answer: 629

   To solve this, rewrite it in the exponential form, $5^2 = \sqrt{x - 4}$. Now it's an equation that can be solved by squaring both sides: $625 = x - 4$. Solving for x, add 4 to each side.

5. Solve for x in $e^{6x-1} = 3$.

   answer: $\dfrac{1 + \ln 3}{6}$

   This needs to be rewritten as a logarithm in base e. It reads $\ln 3 = 6x - 1$. Now, to solve for x, add 1 to each side and divide by 6.

Work Problems

1. Solve for x in $5^{x-2} = 25^{2-x}$.

2. Solve for x in $32^{x+1} = 8^{x^2+1}$.

3. Solve for x in $\log_3 x + \log_3(x - 2) = \log_3 15$.

4. Solve for x in $\log(x + 3) = 3$.

5. Solve for x in $e^{2x-4} = 6$.

Worked Solutions

1. **2** Rewrite the 25 as a power of 5 and then multiply the exponents on that term: $5^{x-2} = \left(5^2\right)^{2-x} = 5^{4-2x}$. Set the exponents equal to one another to get $x - 2 = 4 - 2x$. Solving for x, $3x = 6$, $x = 2$.

2. **2, $-\frac{1}{3}$** Rewrite both the 32 and 8 as powers of 2. Then multiply their exponents: $\left(2^5\right)^{x+1} = \left(2^3\right)^{x^2+1}$, $2^{5x+5} = 2^{3x^2+3}$. Set the exponents equal to one another to get a quadratic equation that can be solved: $5x + 5 = 3x^2 + 3$, $0 = 3x^2 - 5x - 2$. Factor the quadratic: $(3x + 1)(x - 2) = 0$. The two solutions are $-\frac{1}{3}$ and 2.

3. **5** Use the law of logarithms involving the log of a product. This is the reverse, changing a sum of two logs into a product: $\log_3 x(x - 2) = \log_3 15$. Now set the expressions in the logs equal to one another to get $x(x - 2) = 15$ or $x^2 - 2x - 15 = 0$. This factors into $(x - 5)(x + 3) = 0$. There are two solutions to this quadratic equation, 5 and -3. The -3 is extraneous—it doesn't work in the log equation, because it makes two of the log inputs negative numbers.

4. **997** Rewrite this as an exponential expression, $10^3 = x + 3$. Then solve for x in $1000 = x + 3$.

5. **$\frac{4 + \ln 6}{2}$** Rewrite this as a log expression in base e, $\ln 6 = 2x - 4$. To solve for x, add 4 to each side and divide by 2.

Solving Exponential and Logarithmic Inverses

In Chapter 3, you are introduced to the inverse of function $f(x)$, denoted $f^{-1}(x)$, for which the composition of the function yields: $f(x) \circ f^{-1}(x) = f^{-1}(x) \circ f(x) = x$. Not all functions have inverses. But those functions that do have inverses hold to the composition property.

To solve for inverses of exponential and logarithmic functions, you need four properties or rules involving those functions:

1. $b^y = x \Leftrightarrow \log_b x = y$ The exponential/logarithmic equivalence
2. If $\log_b x = \log_b y$, then $x = y$.   Equality property
3. $\log_b x^n = n \log_b x$ The log of a power property
4. $\log_b b = 1$ The log base b of b

The general process for finding the inverse of a function is found in Chapter 3 and is:

1. Rewrite the function replacing $f(x)$ with y.
2. Exchange every x with a y and every y with an x.
3. Solve for y.
4. Rewrite the function replacing the y with $f^{-1}(x)$.

Example Problems

1. Find the inverse of $f(x) = \log_3(x^3 + 1)$.

   1. Rewriting: $y = \log_3(x^3 + 1)$

   2. Exchanging: $x = \log_3(y^3 + 1)$

   3. Using the equivalence, you get $3^x = y^3 + 1$. Now solve for y by subtracting 1 from each side of the equation and taking the cube root of each side.

   $$3^x - 1 = y^3$$
   $$\sqrt[3]{3^x - 1} = \sqrt[3]{y^3}$$
   $$\sqrt[3]{3^x - 1} = y$$

   4. Rewriting: $f^{-1}(x) = \sqrt[3]{3^x - 1}$

2. Find the inverse of $f(x) = 2^{x+4}$.

   1. Rewriting: $y = 2^{x+4}$

   2. Exchanging: $x = 2^{y+4}$

   3. Using the equality property, find the log base 2 of each side.

   $$\log_2 x = \log_2\left(2^{y+4}\right)$$

   Now use the log of a power property and, then, the log base b of b property.

   $$\log_2 x = \log_2\left(2^{y+4}\right)$$
   $$\log_2 x = (y + 4)\log_2 2$$
   $$\log_2 x = (y + 4)(1)$$

   Finally, solve for y.

   $$\log_2 x = y + 4$$
   $$-4 + \log_2 x = y$$

   4. Rewriting: $f^{-1}(x) = -4 + \log_2 x$

Work Problems

Find the inverse of each function.

1. $f(x) = \log_5(x + 7)$
2. $f(x) = \ln(x^5 - 3)$
3. $f(x) = 10^{x-3}$
4. $f(x) = 3e^{x+1}$
5. $f(x) = 4^{2x-3}$

Worked Solutions

1. $f(x) = \log_5(x+7)$

   1. Rewriting: $y = \log_5(x+7)$

   2. Exchanging: $x = \log_5(y+7)$

   3. Using the equivalence, you get $5^x = y + 7$. Subtract 7 from each side of the equation, giving you: $5^x - 7 = y$

   4. Rewriting: $f^{-1}(x) = 5^x - 7$

2. $f(x) = \ln(x^5 - 3)$

   1. Rewriting: $y = \ln(x^5 - 3)$

   2. Exchanging: $x = \ln(y^5 - 3)$

   3. The natural logarithm, denoted ln, has a base of e. Using the equivalence, you get $e^x = y^5 - 3$. Adding 3 to each side of the equation, and then finding the fifth root of each side, you get:

   $$e^x + 3 = y^5$$
   $$\sqrt[5]{e^x + 3} = \sqrt[5]{y^5}$$
   $$\sqrt[5]{e^x + 3} = y$$

   4. Rewriting: $f^{-1}(x) = \sqrt[5]{e^x + 3}$

3. $f(x) = 10^{x-3}$

   1. Rewriting: $y = 10^{x-3}$

   2. Exchanging: $x = 10^{y-3}$

   3. Using the equality property, find the log base 10 of each side.

   $$\log_{10} x = \log_{10}\left(10^{y-3}\right)$$

   Now use the log of a power property and, then, the log base b of b property.

   $$\log_{10} x = \log_{10}\left(10^{y-3}\right)$$
   $$\log_{10} x = (y-3)\log_{10} 10$$
   $$\log_{10} x = (y-3)(1)$$

   Finally, solve for y. Also, logs base 10 are the *common logarithms* and can be written log, without the base shown.

   $$\log_{10} x = y - 3$$
   $$3 + \log_{10} x = y$$
   $$3 + \log x = y$$

   4. Rewriting: $f^{-1}(x) = 3 + \log x$

4. $f(x) = 3e^{x+1}$

   1. Rewriting: $y = 3e^{x+1}$

   2. Exchanging: $x = 3e^{y+1}$

3. First, divide each side of the equation by 3. Then, using the equality property, find ln of each side.

$$x = 3e^{y+1}$$

$$\frac{x}{3} = e^{y+1}$$

$$\ln\left(\frac{x}{3}\right) = \ln\left(e^{y+1}\right)$$

The natural log, ln, is base e. So, using the log of a power property and, then, the log base b of b property,

$$\ln\left(\frac{x}{3}\right) = \ln\left(e^{y+1}\right)$$

$$\ln\left(\frac{x}{3}\right) = (y+1)\ln e$$

$$\ln\left(\frac{x}{3}\right) = (y+1)(1)$$

Finally, solve for y.

$$\ln\left(\frac{x}{3}\right) = y+1$$

$$-1 + \ln\left(\frac{x}{3}\right) = y$$

4. Rewriting: $f^{-1}(x) = -1 + \ln\left(\frac{x}{3}\right)$

5. $f(x) = 4^{2x-3}$

1. Rewriting: $y = 4^{2x-3}$

2. Exchanging: $x = 4^{2y-3}$

3. Using the equality property, find the log base 4 of each side.

$$\log_4 x = \log_4\left(4^{2y-3}\right)$$

Now use the log of a power property and, then, the log base b of b property.

$$\log_4 x = \log_4\left(4^{2y-3}\right)$$

$$\log_4 x = (2y-3)\log_4 4$$

$$\log_4 x = (2y-3)(1)$$

Finally, solve for y.

$$\log_4 x = 2y - 3$$

$$3 + \log_4 x = 2y$$

$$\frac{3 + \log_4 x}{2} = y$$

4. Rewriting: $f^{-1}(x) = \dfrac{3 + \log_4 x}{2} = \dfrac{3}{2} + \dfrac{1}{2}\log_4 x$

Chapter 6

Graphing

Graphing a function is an effective way of displaying its characteristics and allowing you to draw conclusions about it and what it might represent. Graphs can be done with point-by-point plots, or a more effective method is to take advantage of quickly determined information and general knowledge about the type of function and put that to use. Several of the more common types of graphs are discussed in this chapter, followed by a section on applying transformations to the graphs.

Graphing Polynomials

A polynomial is a function whose graph is a smooth curve and that has a domain of all real numbers. (For more on Polynomials, see Chapter 4.) The input values (domain values) go from negative infinity (very small) to positive infinity (very large), and the output values (range values) eventually become infinitely large and/or infinitely small. It's the intercepts, turning points, and local extreme values that are usually the most interesting characteristics of polynomials.

Consider the general polynomial $f(x) = a_n x^n + a_{n-1} x^{n-1} + a_{n-2} x^{n-2} + \ldots + a_1 x^1 + a_0$ where n is a positive integer or 0, and the a's are real numbers. A lot can be learned from this equation when there are actual numbers replacing the n's and the a's.

1. If a_n is positive, then the graph of the polynomial goes infinitely *high* as x's (the input values) get very large. If a_n is negative, then the graph goes infinitely *low* as the x's get very large.

2. If n is an even number, then the graph of the polynomial goes infinitely high or low as x gets both very large and very small, depending on whether a_n is positive or negative.

3. If n is odd, then there must be at least one x-intercept.

4. The maximum number of x-intercepts is n.

5. The maximum number of turning points (where the graph switches from going up to going down or vice versa) in the graph is $n - 1$.

6. The y-intercept of the graph is $(0, a_0)$.

7. The x-intercepts are determined by setting $f(x) = 0$ and solving for x.

8. If x is where there is an x-intercept (zero, root, solution of $f(x) = 0$), and if it is a double root or any multiple of 2 as a root, then the graph of the polynomial doesn't cross at that intercept. It does a "touch and go."

9. If all of the powers of x are even (a constant is considered to have an even power, it's the coefficient of the even power, x^0), then the graph is symmetric about the y-axis. There's a mirror image of the graph across that axis.

Example Problems

These problems show answers and solutions.

1. Graph $y = x^4 - 2x^3 - 7x^2 + 8x + 12$.

   answer:

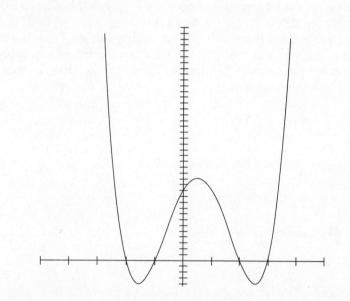

The lead coefficient (the coefficient of the highest power, x^4) is positive, so as x gets infinitely large or infinitely small, the y values will get infinitely large. Because the highest power is 4, the graph has at most four x-intercepts and at most three turning points. The y-intercept is at (0,12). Set $y = 0$ and solve $x^4 - 2x^3 - 7x^2 + 8x + 12 = 0$. See Chapter 4 on "Solving Polynomial Equations" for more on this. The solutions are $x = -2, -1, 2, 3$, so the x-intercepts are $(-2,0)$, $(-1, 0)$, $(2,0)$, $(3,0)$. The graph will be coming downward from the left of $(-2,0)$ and upward to the right of $(3,0)$. It crosses at $(-1,0)$ and $(2,0)$. Since it crosses the x-axis four times, there'll be three turning points. A local minimum occurs between -2 and -1; a local maximum occurs between -1 and 2; and another local minimum occurs between 2 and 3.

2. Graph $y = -x^5 + 10x^4 - 24x^3 - 10x^2 + 25x$.

   answer:

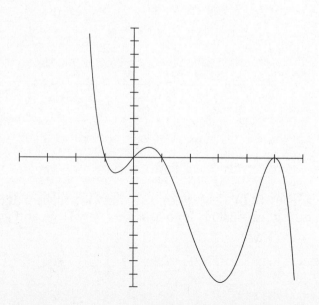

The lead coefficient is negative, so as x becomes infinitely large, the y values get very small and the graph goes downward. Because the highest power is 5, an odd number, the two "ends" of the graph will go in opposite directions as far as up and down. So the y values will get very large as x gets smaller and smaller. There are at most five x-intercepts and at most four turning points. The constant, a_0 is equal to 0, so the y-intercept is (0,0). That's going to be an x-intercept, too. Solve $y = 0$ to get the x-intercepts. When $-x^5 + 10x^4 - 24x^3 - 10x^2 + 25x = 0$, the solutions are when $x = -1, 0, 1, 5, 5$. There's a double root at 5, so the x-intercept there will have a "touch and go." The x-intercepts are $(-1,0)$, $(0,0)$, $(1,0)$, $(5,0)$. The graph comes downward from the left of -1, crosses there, goes up through $(0,0)$, comes back down through $(1,0)$, comes up, does a "touch and go" at $(5,0)$, and then goes down infinitely low.

Work Problems
Use these problems to give yourself additional practice.

1.   Graph $y = x^3 - 30x^2 - 18x + 40$.

2.   Graph $y = x^4 + 4x^3 - 13x^2 - 28x + 60$.

3.   Graph $y = -x^4 - 4x^3 + 13x^2 + 4x - 12$.

4.   Graph $y = x^6 - 30x^4 + 129x^2 - 100$.

5.   Graph $y = -x^5 + x^4 + 16x^3 - 16x^2$.

Worked Solutions

1.

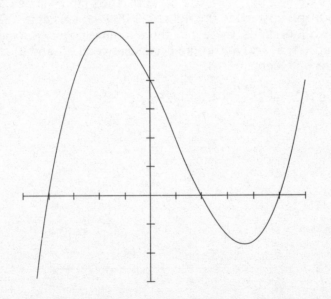

The x-intercepts are at $(-4,0)$, $(2,0)$, $(5,0)$, and the y-intercept is at $(0,40)$. The graph has two turning points. It goes infinitely high as x gets very large and goes infinitely low as x gets very small.

2.

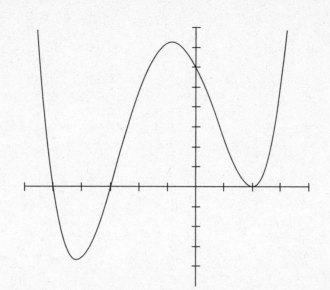

The x-intercepts are at (−5,0), (−3,0), (2,0) with a double root or "touch and go" at (2,0). The y-intercept is at (0,60). The lead coefficient is positive and the highest power is even, so the graph goes infinitely high both as x gets very large and very small.

3.

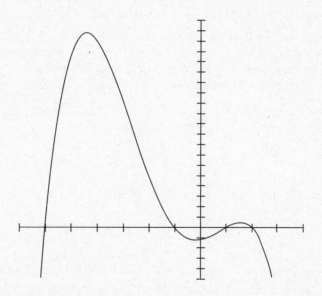

The x-intercepts are at (−6,0), (−1,0), (1,0), (2,0), and the y-intercept is at (0,−12). The lead coefficient is negative, and the highest power is even, so the graph goes infinitely low both as x gets very large and very small.

4.

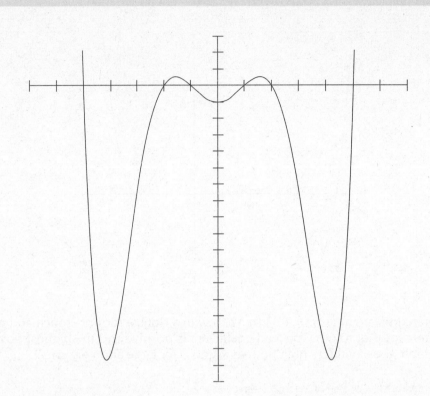

The *x*-intercepts are at (−5,0), (−2, 0), (−1, 0), (1, 0), (2, 0), (5, 0), and the *y*-intercept is at (0, −100). The lead coefficient is positive, and the highest power is even, so the graph gets infinitely large as *x* gets very large and very small. All of the powers are even, so the graph of this polynomial is symmetric about the *y*-axis. The two sides will be mirror images of one another.

5.

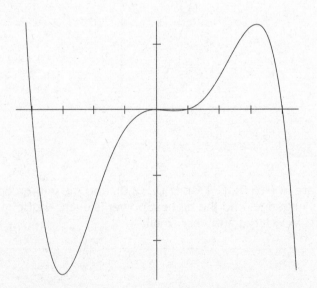

The *x*-intercepts are at (−4, 0), (0,0), (1,0), (4,0), and the *y*-intercept is also at (0,0). The solution to the equation, $-x^5 + x^4 + 16x^3 - 16x^2 = 0$, has a double root when *x* = 0, so the graph doesn't cross the axis at that point. The lead coefficient is negative, so the graph will go infinitely low as *x* gets larger. The highest power is odd, so the *y*-values will get infinitely large as *x* gets infinitely small.

Graphing Rational Functions

A rational function is one that has the form $y = \dfrac{f(x)}{g(x)}$ where $f(x)$ and $g(x)$ are both polynomials.

Things that make rational functions so special are their restrictions on the domain and their asymptotes. The domain is often restricted because of the fraction. If input values, x, result in a 0 in the denominator, then the function has no y value for those values of x. The domain cannot include those values. If $\dfrac{f(x)}{g(x)}$ is in lowest terms—no factors in common, then any x's that make the denominator equal to 0 can be shown graphically with a *vertical asymptote*. A vertical asymptote is an aid to graphing. It's drawn in with a dashed line to show that it isn't a part of the graph of the function. The graph of the function will get very close to the vertical asymptote, hugging it as it moves up to infinitely high values or down to infinitely low values. The graph will never cross a vertical asymptote, just get very close to it. The equation of a vertical asymptote is $x = h$. The h is the value of x that makes the denominator equal to 0. To solve for vertical asymptotes, set $g(x) = 0$ and solve for the values of x that make this happen.

Another type of asymptote is a *horizontal asymptote*. This is another graphing aid. The equation of a horizontal asymptote is $y = k$. The k is the number that the y or output values get very close to as x gets infinitely large or infinitely small. This asymptote is also graphed with a dashed line to show that it isn't a part of the graph of the function. A function might cross a horizontal asymptote at some point, but it hugs the horizontal asymptote as you move to the far right or far left.

To solve for the equation of the horizontal asymptote, the following multi-part rule can be used. Solving for a horizontal asymptote is done more formally in mathematics courses where limits are discussed, but, for now, this method works fine. If the rational function can be written $y = \dfrac{f(x)}{g(x)} = \dfrac{a_n x^n + a_{n-1} x^{n-1} + \ldots}{b_m x^m + b_{m-1} x^{m-1} + \ldots}$, then a comparison of the values of n and m will determine the equation of the horizontal asymptote.

If $n > m$, then there is no horizontal asymptote.

If $n < m$, then the horizontal asymptote is $y = 0$, the x-axis.

If $n = m$, then the horizontal asymptote is $y = \dfrac{a_n}{b_m}$, the fraction made from the two lead coefficients.

Example Problems

These problems show the answers and solutions.

1. Graph $y = \dfrac{4x - 1}{x - 3}$.

answer:

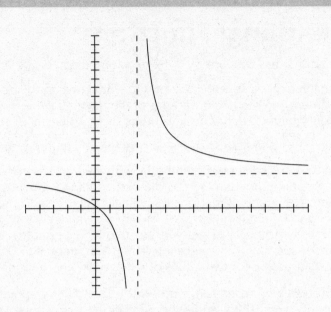

The vertical asymptote is $x = 3$, because a 3 in place of the variable will make the denominator equal to 0. The horizontal asymptote is $y = 4$, because the highest powers of the numerator and denominator are both 1, and the fraction made by the coefficients of those terms is $\frac{4}{1}$. The x-intercept is $\left(\frac{1}{4}, 0\right)$, because when y is equal to 0, the value of x that makes the numerator of the fraction equal to 0 is $\frac{1}{4}$. The y-intercept is equal to $\frac{1}{3}$, because when x is equal to 0, $y = \frac{-1}{-3}$. Plot a few points to determine where the graph lies, having the graph of the function hug the two asymptotes to the left and right, up and down.

2.   Graph $y = \dfrac{x - 6}{x^2 - x + 2}$.

answer:

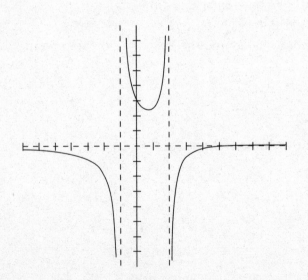

There are two vertical asymptotes, because $x^2 - x - 2 = 0$ when $x = -1$ and $x = 2$. The horizontal asymptote is $y = 0$. The x-intercept is $(6, 0)$, and the y-intercept is $(0, 3)$. Plot a few points to determine where the graph lies.

Work Problems

Use these problems to give yourself additional practice.

1. Graph $y = \dfrac{x}{x+2}$.

2. Graph $y = \dfrac{x^2-1}{x^2-9}$.

3. Graph $y = \dfrac{x-4}{x^2+1}$.

4. Graph $y = \dfrac{5x-5}{x+1}$.

5. Graph $y = \dfrac{2x-5}{x^2-4x-5}$.

Worked Solutions

1.

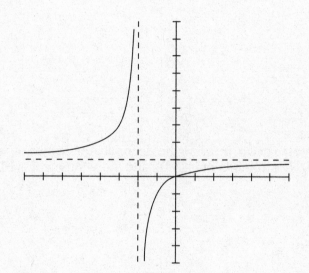

The vertical asymptote is $x = -2$, and the horizontal asymptote is $y = 1$. The only intercept is $(0,0)$.

2.

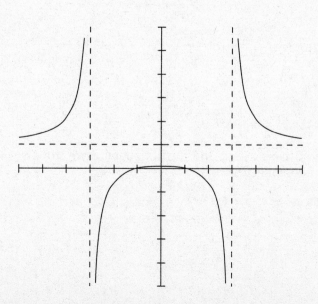

The two vertical asymptotes are at $x = 3$ and $x = -3$. The horizontal asymptote is at $y = 1$. The two x-intercepts are found by setting y equal to 0 and solving $x^2 - 1 = 0$. They are $(1,0)$ and $(-1,0)$. The y-intercept is found by setting x equal to 0. It's $\left(0, \frac{1}{9}\right)$. All of the powers on the variables are even numbers (constants have a variable with power 0), so this graph is symmetric about the y-axis.

3.

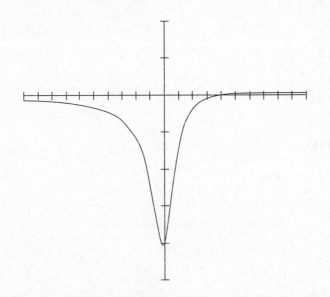

No vertical asymptotes occur, because the denominator cannot equal 0. The horizontal asymptote is $y = 0$. The x-intercept is $(4,0)$, and the y-intercept is $(0,-4)$.

4.

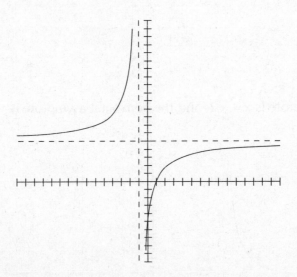

The vertical asymptote is $x = -1$, and the horizontal asymptote is $y = 5$. The x-intercept is $(1,0)$, and the y-intercept is $(0,-5)$.

5.

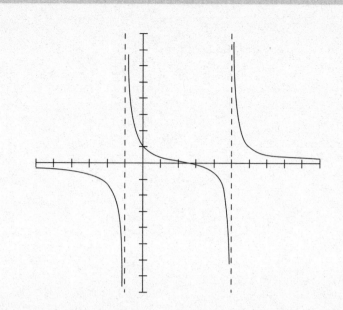

The two vertical asymptotes are found by setting the denominator equal to 0. $x^2 - 4x - 5 = 0$ when $x = 5$ and $x = -1$. The horizontal asymptote is $y = 0$. The x-intercept is $\left(\frac{5}{2}, 0\right)$, and the y-intercept is $(0, 1)$.

Graphing Radical Functions

A radical function can have a domain of all real numbers or a domain that is restricted. The domain can be all real numbers if the root of the radical is an odd number, such as in $y = \sqrt[3]{x}$ or $y = \sqrt[5]{x - 2}$. The domain is restricted to values of x that make the expression under the radical positive or 0 if the root is even, such as in $y = \sqrt{x - 4}$ or $y = \sqrt[4]{1 - x}$. In both cases, though, the value or values that make the expression under the radical equal to 0 are where the graph of the function becomes nearly vertical. The reason that this happens is explained in a calculus course—something to look forward to! For now, though, just use that information to help with the graphs.

Example Problems

These problems show the answers and the solutions.

1. Graph $y = \sqrt{x}$.

   answer:

   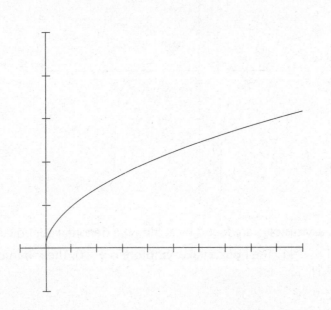

   The domain is all positive numbers and 0, so the graph will all be to the right of the *y*-axis. The range is also all positive numbers and 0, so the graph will be in the first quadrant only. The graph becomes nearly vertical when *x* = 0.

2. Graph $y = \sqrt[3]{x}$.

   answer:

   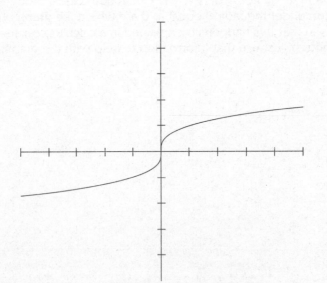

   The domain and range are both all real numbers. The graph becomes nearly vertical when *x* = 0.

3. Graph $y = \sqrt[4]{x-1}$.

answer:

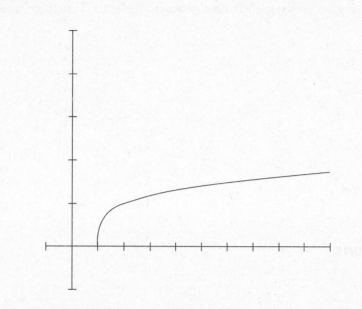

The domain includes all numbers equal to 1 and larger. The range is all positive numbers and 0. The graph becomes nearly vertical at $x = 1$.

4. Graph $y = \sqrt{4 - x^2}$.

answer:

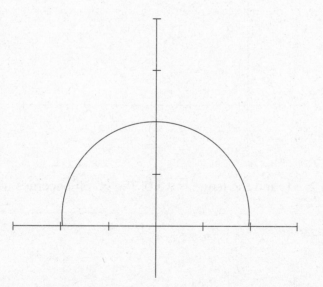

The domain of this function is $-2 \leq x \leq 2$, because anything to the left of -2 or to the right of 2 will result in a negative under the radical. The range is $0 \leq y \leq 2$. Functions with this form, $f(x) = \sqrt{a^2 - x^2}$, have graphs that are semicircles with a radius of a. The graph becomes nearly vertical at each end of the domain.

Work Problems

Use these problems to give yourself additional practice. Refer to "Function Transformations" in Chapter 3.

1. Graph $y = \sqrt{x + 1}$.

2. Graph $y = \sqrt[4]{9 - x}$.

3. Graph $y = \sqrt[3]{x + 2}$.

4. Graph $y = \sqrt[5]{x}$.

5. Graph $y = \sqrt{1 - x^2}$.

Worked Solutions

1.

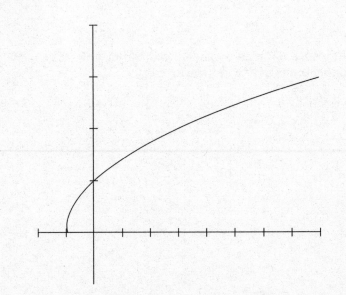

The domain is $x \geq -1$, and the range is $y \geq 0$. The graph becomes nearly vertical when $x = -1$.

2.

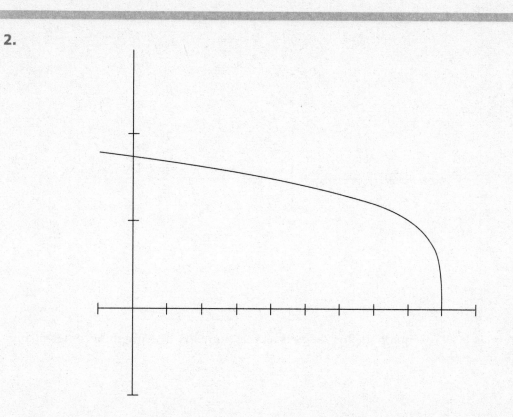

The domain is $x \leq 9$; the range is $y \geq 0$. The graph becomes nearly vertical when $x = 9$.

3.

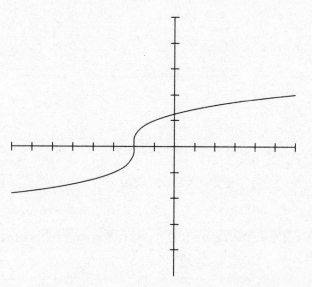

The domain is all real numbers; the range is all real numbers. The graph becomes nearly vertical at $x = -2$.

4.

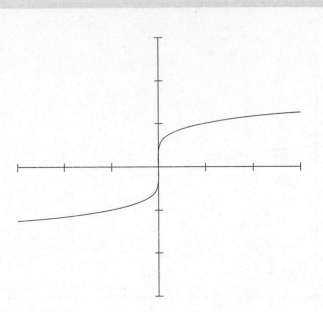

The domain is all real numbers; the range is all real numbers. The graph becomes nearly vertical at $x = 0$.

5.

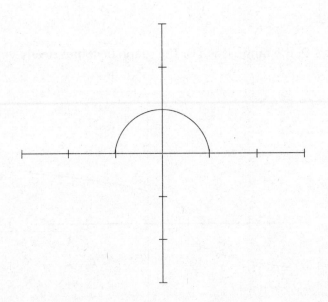

The domain is $-1 \leq x \leq 1$; the range is $0 \leq y \leq 1$. The graph becomes nearly vertical at $x = -1$ and $x = 1$.

Graphing Absolute Value and Logarithmic Functions

Absolute value functions of the general form $f(x) = |ax + b|$ have distinctive shapes. They are V-shaped and can be graphed quickly if you know how to read the equation. The graph of the function $y = |x|$ has its lowest point at the origin and diagonals in the first and second quadrants. The part of the graph in the first quadrant has the equation $y = x$, and the part of the graph in the second quadrant has the equation $y = -x$.

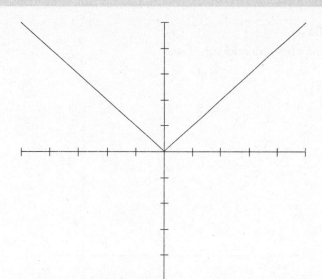

In the general form, $f(x) = |ax + b|$, the bottom of the V is at the point $\left(-\frac{b}{a}, 0\right)$. The steepness of the V is determined by the value of a. The a acts as the slope of the right side, and $-a$ is the slope of the left side. There are other variations, raising it off the x-axis and flipping it. These are covered with all of the transformations in the next section of this chapter.

Logarithmic functions are the inverses of exponential functions. For more on this, see "Logarithmic Functions" in Chapter 5. Their graphs have many similarities to exponential functions. This is because, when functions are inverses of one another, their graphs are symmetric (mirror images) about the line $y = x$. For instance, the function $y = 2^x$ has a graph with a y-intercept of $(0,1)$ and a *horizontal* asymptote of $y = 0$ to the left of this intercept. The inverse of this function, $y = \log_2 x$ has an x-intercept of $(1,0)$ (notice that the coordinates are reversed). This inverse has a *vertical* asymptote of $x = 0$ below this intercept. Look at the graphs of the two functions that follow.

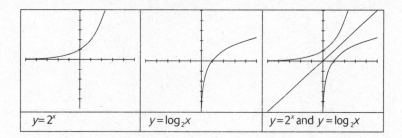

Example Problems

These problems show the answers and solutions.

1. Graph $y = |x + 3|$.

    answer:

    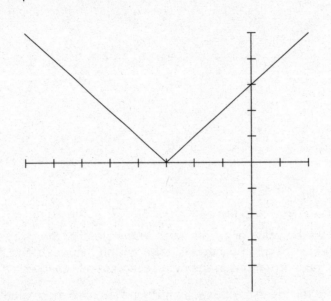

    The bottom of the V is at $\left(-\frac{b}{a}, 0\right)$, which, in this case, is $\left(-\frac{3}{1}, 0\right) = (-3, 0)$. The value of a is 1, so the slope of the right side of the graph is 1, and the slope of the left side of the graph is -1.

2. Graph $y = |2x - 8|$.

    answer:

    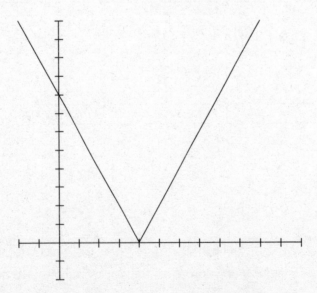

    The lowest point of the V is at $\left(-\frac{b}{a}, 0\right) = \left(-\frac{-8}{2}, 0\right) = (4, 0)$. The slope of the right side of the graph is 2, and the slope of the left side of the graph is -2.

3. Graph $y = \left| \frac{1}{3}x + 2 \right|$.

   answer:

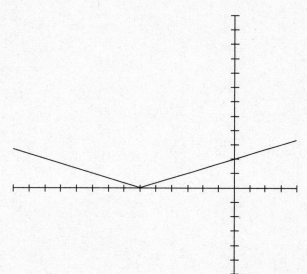

The lowest part of the V is at $\left(-\frac{2}{1/3}, 0 \right) = (-6, 0)$. The slope of the right side is $\frac{1}{3}$, and the slope of the left side is $-\frac{1}{3}$.

4. Graph $y = \log_3 x$.

   answer:

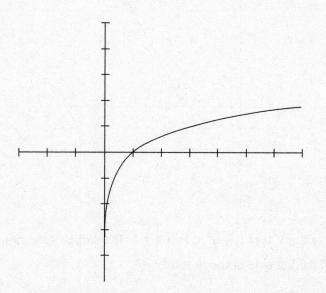

The x-intercept is $(1, 0)$. Some of the other points on the graph are $\left(\frac{1}{3}, -1 \right)$, $(3, 1)$, $(9, 2)$. The y-axis, $x = 0$, is a vertical asymptote for the graph in the fourth quadrant.

Work Problems

Use these problems to give yourself more practice.

1.   Graph $y = |2x - 3|$.

2.   Graph $y = \left|\frac{1}{4}x + 2\right|$.

3.   Graph $y = |2 - x|$.

4.   Graph $y = \log_4 x$.

5.   Graph $y = \log_{1/2} x$.

Worked Solutions

1.   **answer:**

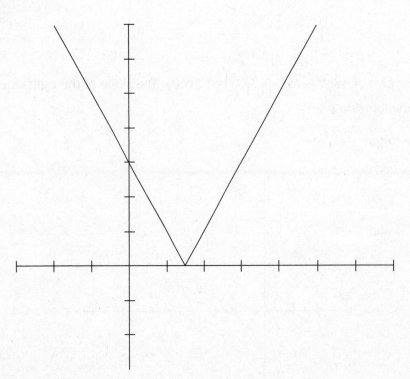

The lowest part of the V is at $\left(-\frac{-3}{2}, 0\right) = \left(\frac{3}{2}, 0\right)$. The slope of the right side of the graph is 2, and the slope of the left side of the graph is -2.

2. **answer:**

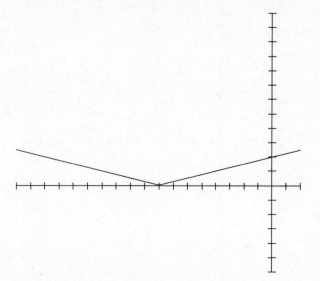

The lowest part of the V is at $\left(-\dfrac{2}{1/4}, 0\right) = (-8, 0)$. The slope of the right side of the graph is $\dfrac{1}{4}$, and the slope of the left side of the graph is $-\dfrac{1}{4}$.

3. **answer:**

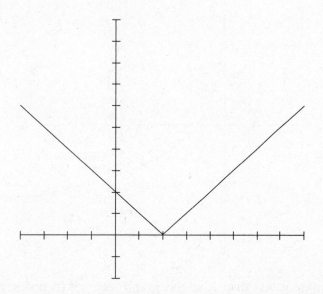

The lowest part of the V is at $\left(-\dfrac{2}{-1}, 0\right) = (2, 0)$. The slope of the right side of the graph is 1, and the slope of the left side of the graph is -1. Even though the value of a is -1, the function of absolute value allows that $|2 - x| = |x - 2|$, so the slopes can be interpreted either way.

4. **answer:**

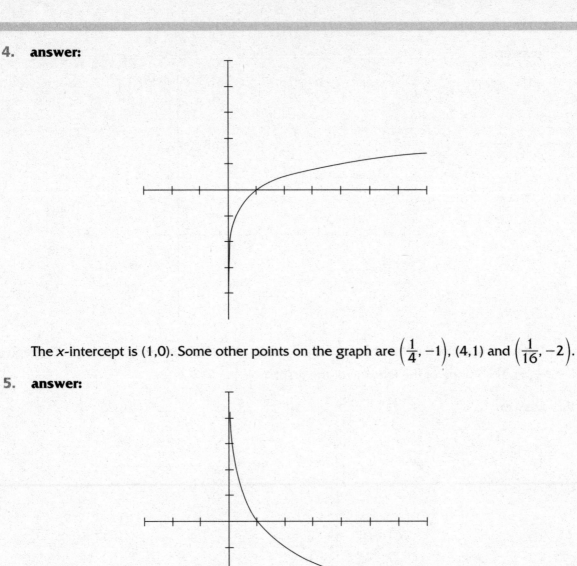

The x-intercept is (1,0). Some other points on the graph are $\left(\frac{1}{4}, -1\right)$, (4,1) and $\left(\frac{1}{16}, -2\right)$.

5. **answer:**

The value of the base is less than 1, so this graph reverses its properties just like what happens between the graphs of the exponential counterparts, $y = 2^x$ and $y = \left(\frac{1}{2}\right)^x$. For more on this, see "Exponential Functions" in Chapter 5. The x-intercept is still at (1,0), but the graph has its vertical asymptote above that intercept, instead of below it. Some other points on the graph are $\left(\frac{1}{2}, 1\right)$, $\left(\frac{1}{4}, 2\right)$, and (2,−1).

Graphing Functions Using Transformations

Some function transformations are the slides (translations), flips (reflections), and warps that change the basic function to one very much like the original but with predictable adjustments. For more details on how they work, see Chapter 3 on "Function Transformations." The advantage to using function transformations when graphing is that you can quickly sketch the graph of many different functions by just applying the transformations to the basic forms.

The function transformations (assume that c and k are positive) used here are

Slides (translations)	$f(x) \pm c$	Slides the graph up (+c) or down (−c) units
	$f(x \pm c)$	Slides the graph right (−c) or left (+c) units
Flips (reflections)	$-f(x)$	Flips the graph over a horizontal line
	$f(-x)$	Flips the graph over a vertical line
Warps ($k > 1$)	$kf(x)$	Makes the graph steeper
	$\frac{1}{k}f(x)$	Makes the graph flatter

Some of the basic functions and their graphs are shown here.

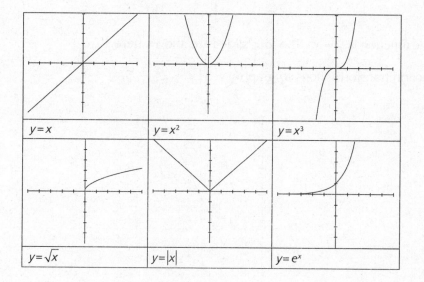

Example Problems

These problems show the answers and solutions.

1. Use function transformations to graph $y = x^2 + 3$, $y = x^2 - 3$, $y = (x + 3)^2$, and $y = (x - 3)^2$.

    answer:

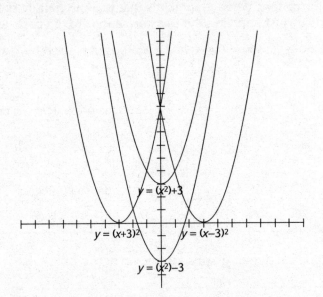

    The basic function is $y = x^2$. The four slides are shown here.

2. Use function transformations to graph $y = 3|x - 2| + 5$.

    answer:

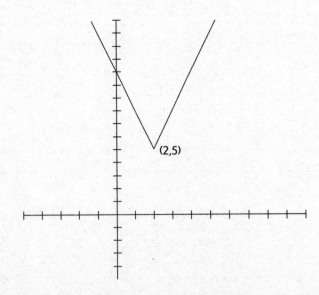

    The basic graph here is $y = |x|$. The bottom of the V has been slid 2 units to the right and 5 units up. The graph is steeper, because the 3 makes the slopes of the two half lines 3 and -3.

3. Use function transformations to graph the function $y = -\frac{1}{2}\sqrt{x+4}$.

    answer:

    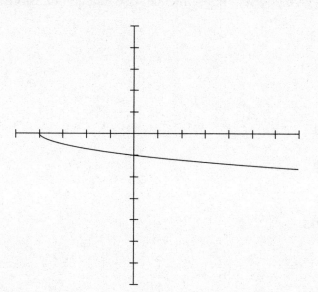

    The basic graph here is $y = \sqrt{x}$. The negative multiplier flipped the basic graph over the horizontal x-axis. The multiplier of $\frac{1}{2}$ made it flatter (less steep), and the +4 under the radical slid the graph 4 units to the left.

4. Use function transformations to graph $y = 2e^{x-1} - 3$.

    answer:

    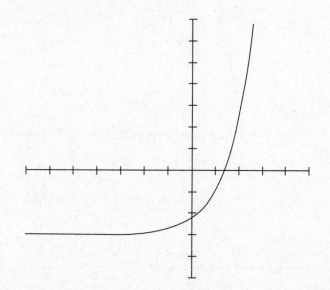

    The basic graph is $y = e^x$. The 2 multiplier makes it steeper. The slide $x - 1$ shifts it 1 unit to the right, and the slide -3 moves it 3 units down—making its horizontal asymptote $y = -3$.

Work Problems

Use these problems to give yourself more practice. Use function transformations to graph each.

1. $y = 2(x + 1)^3 - 3$

2. $y = -3(x - 4)^2 + 1$

3. $y = 5x + 2$

4. $y = \frac{1}{4}|x + 3| + 4$

5. $y = -\sqrt{2 - x} + 5$

Worked Solutions

1.

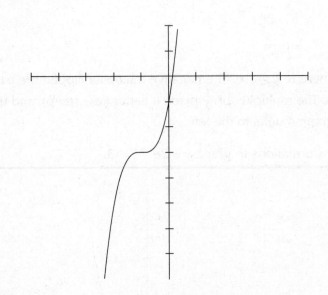

The basic graph is $y = x^3$. The 2 multiplier makes it steeper. The slides are 1 unit to the left and 3 units down.

2.

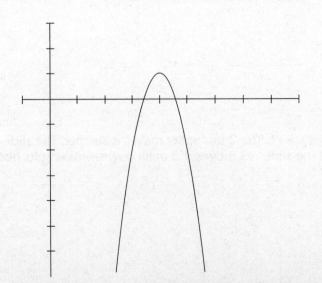

The basic graph is $y = x^2$. The negative multiplier flips the curve over a horizontal line, making it open downward instead of upward. The 3 multiplier makes it steeper. The slides are 4 units to the right and 1 unit up.

3.

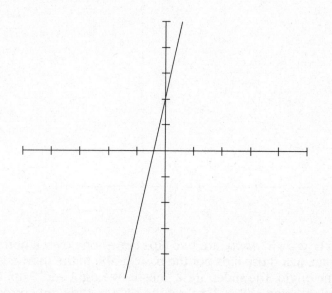

The basic graph is $y = x$. The 5 multiplier makes it steeper, and the slide is 2 units up. Of course, you could have easily drawn this using the properties of a line with a slope of 5 and y-intercept of 2.

4.

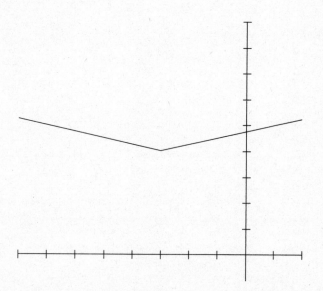

The basic graph is $y = |x|$. The $\frac{1}{4}$ multiplier makes it much flatter. The slides are 3 units to the left and 4 units up.

5.

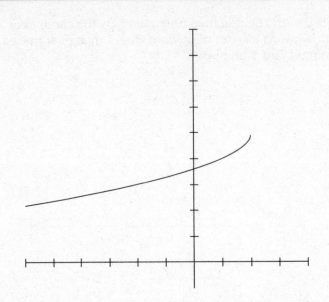

The basic graph is $y = \sqrt{x}$. There are two flips here—one over a horizontal line and one over a vertical line. Just those flips put the basic graph in the third quadrant with its "endpoint" at the origin. The slides are 2 units to the right and 5 up. Think of the value under the radical as being $-(x - 2)$ to get the slide to the right correctly drawn. The function then is written $y = -\sqrt{-(x - 2)} + 5$.

Chapter 7
Other Equations

In this chapter, you'll find solutions to quadratic and radical equations. These equations are also covered in Chapter 2, but in this chapter we handle solutions that are not real. First, you'll find information on complex numbers—imaginary numbers. Also in this chapter is a discussion of rational equations and variation. It's a sort of wrap-up of the situations you'll face when solving algebraic equations.

Radical Equations and Conjugates

Solving simple radical equations is covered in Chapter 2. What's new here are two situations that can arise when solving radical equations. One situation involves equations that are fractions with radicals in the denominator. The process used to simplify these fractions is called *rationalizing* the denominator. The other situation arises when squaring both sides of an equation to solve it and finding that you have to repeat that process—square both sides again.

Rationalizing a Denominator

The expressions $\frac{x}{\sqrt{2}}$, $\frac{5}{\sqrt{x}}$, $\frac{x}{2+\sqrt{3}}$, and $\frac{2}{\sqrt{x-1}}$ are all fractions with square roots in the denominator. It isn't good form to leave a square root in the denominator, because you could be dividing by an irrational number. When there's one term in the denominator that's a square root, multiplying both numerator and denominator by that square root (in essence, multiplying by 1) will *rationalize* the denominator. If two terms are in the denominator, one or both a square root, then multiplying both the numerator and denominator by the *conjugate* of that denominator will *rationalize* the denominator. The *conjugate* of the *sum or difference* of two terms is the *difference or sum* of the same two terms.

Example Problems

These problems show the answers and solutions.

1. Rationalize $\frac{x}{\sqrt{2}}$.

   answer: $\frac{x\sqrt{2}}{2}$

   Multiply the numerator and denominator of the original fraction by $1 = \frac{\sqrt{2}}{\sqrt{2}}$:

   $\frac{x}{\sqrt{2}} \cdot \frac{\sqrt{2}}{\sqrt{2}} = \frac{x\sqrt{2}}{\sqrt{4}} = \frac{x\sqrt{2}}{2}$. This may not look as nice and neat as the original fraction, but it is, in fact, a simpler expression.

2. Rationalize $\dfrac{5}{\sqrt{x}}$.

   answer: $\dfrac{5\sqrt{x}}{x}$

   Multiply the numerator and denominator by $1 = \dfrac{\sqrt{x}}{\sqrt{x}}$: $\dfrac{5}{\sqrt{x}} \cdot \dfrac{\sqrt{x}}{\sqrt{x}} = \dfrac{5\sqrt{x}}{\sqrt{x^2}} = \dfrac{5\sqrt{x}}{|x|}$. Technically, $\sqrt{x^2} = |x|$, but, because there is still $\sqrt{x}$ in the expression, the restriction on the value of x (that it has to be positive) is preserved.

3. Rationalize $\dfrac{x}{2+\sqrt{3}}$.

   answer: $x\left(2 - \sqrt{3}\right)$

   Multiply the numerator and denominator by $1 = \dfrac{2-\sqrt{3}}{2-\sqrt{3}}$. The expression $2 - \sqrt{3}$ is the *conjugate* of $2 + \sqrt{3}$, because they're the same two numbers with the opposite operation between them.

   $$\dfrac{x}{2+\sqrt{3}} \cdot \dfrac{2-\sqrt{3}}{2-\sqrt{3}} = \dfrac{x\left(2-\sqrt{3}\right)}{\left(2+\sqrt{3}\right)\left(2-\sqrt{3}\right)} = \dfrac{x\left(2-\sqrt{3}\right)}{4-\sqrt{9}} = \dfrac{x\left(2-\sqrt{3}\right)}{4-3} = x\left(2-\sqrt{3}\right)$$

4. Rationalize $\dfrac{2}{\sqrt{x}-1}$.

   answer: $\dfrac{2\left(\sqrt{x}+1\right)}{x-1}$

   Multiply the numerator and denominator by the conjugate of the denominator:

   $$\dfrac{2}{\sqrt{x}-1} \cdot \dfrac{\sqrt{x}+1}{\sqrt{x}+1} = \dfrac{2\left(\sqrt{x}+1\right)}{\left(\sqrt{x}-1\right)\left(\sqrt{x}+1\right)} = \dfrac{2\left(\sqrt{x}+1\right)}{\sqrt{x^2}-1} = \dfrac{2\left(\sqrt{x}+1\right)}{|x|-1}$$

   The expression still has $\sqrt{x}$, so $x \geq 0$.

Radical Equations—Squaring More Than Once

Squaring both sides of an equation is an effective way of solving radical equations. The only caution is that this process can produce extraneous solutions. Any solutions obtained from changing the original equation to a solvable form need to be checked.

Example Problems

These problems show the answers and solutions.

1. Solve $\sqrt{x+1} + 2 = \sqrt{3x+1}$.

   answer: $x = 8$

   Even if you moved both radicals to the same side, there's no way to avoid having a radical in the equation after squaring both sides. Just leave the equation the way it is so there's only one radical on each side. Squaring both sides, $\left(\sqrt{x+1}+2\right)^2 = \left(\sqrt{3x+1}\right)^2$, you get the square of an expression with two terms on the left: $\left(\sqrt{x+1}\right)^2 + 2\cdot 2\sqrt{x+1} + 2^2 = 3x+1$; $x+1+4\sqrt{x+1}+4 = 3x+1$. Now move all the terms without a radical to the right side by adding the opposites, $4\sqrt{x+1}=2x-4$. Dividing each term by 2 will make the numbers smaller before squaring both sides again.

$$2\sqrt{x+1}=x-2$$
$$\left(2\sqrt{x+1}\right)^2 = (x-2)^2$$

This is equivalent to $4(x+1)=x^2-4x+4$. Distribute on the left and then move everything to the right to solve the resulting quadratic equation: $4x+4=x^2-4x+4$. $x^2-8x=0$. This quadratic factors and gives you two solutions: $x(x-8)=0$, $x=0$, $x=8$. Substituting those values back into the original equation, if $x=0$, $\sqrt{0+1}+2=\sqrt{0+1}$ giving you $1+2=1$. This is not true, so 0 cannot be a solution of the radical equation—it's extraneous. If $x=8$, $\sqrt{8+1}+2=\sqrt{3(8)+1}$, giving you $3+2=\sqrt{25}=5$. This statement is true, so 8 is the solution.

2. Solve $\sqrt{21-4x}-\sqrt{x+5}=\sqrt{2x+11}$.

answer: $x=-1$

Squaring both sides results in a middle term that has the product of two radicals.

$$\left(\sqrt{21-4x}-\sqrt{x+5}\right)^2 = \left(\sqrt{2x+11}\right)^2; \quad 21-4x-2\sqrt{21-4x}\sqrt{x+5}+x+5=2x+11$$

Move all the terms without radicals to the right by adding their opposites to each side: $-2\sqrt{21-4x}\sqrt{x+5}=5x-15$. Now square both sides again.

$$\left(-2\sqrt{21-4x}\sqrt{x+5}\right)^2 = (5x-15)^2; \quad 4(21-4x)(x+5)=25x^2-150x+225$$

Multiply out the left side: $4(-4x^2+x+105)=-16x^2+4x+420=25x^2-150x+225$. Now move all the terms to the right and solve the quadratic: $41x^2-154x-195=0$. This is not an easy one to factor, but it does. You might resort to the quadratic formula: $41x^2-154x-195=(41x-195)(x+1)=0$. Testing both $x=\frac{195}{41}$ and $x=-1$, only the $x=-1$ works; the other is extraneous.

Work Problems

Use these problems to give yourself more practice.

1. Rationalize $\dfrac{3}{\sqrt{x}+1}$.

2. Rationalize $\dfrac{2}{\sqrt{5}-\sqrt{x}}$.

3. Solve $\sqrt{5x+1}=\sqrt{11-x}+4$.

4. Solve $\sqrt{6x+1}+\sqrt{2x+3}=\sqrt{8x+12}$.

5. Solve $x\sqrt{2}+5=x\sqrt{3}$.

Worked Solutions

1. $\dfrac{3\left(\sqrt{x}-1\right)}{x-1}$ Multiply numerator and denominator by the conjugate of the denominator:

$$\dfrac{3}{\sqrt{x}+1}\cdot\dfrac{\sqrt{x}-1}{\sqrt{x}-1}=\dfrac{3\left(\sqrt{x}-1\right)}{\sqrt{x^2}-1}=\dfrac{3\left(\sqrt{x}-1\right)}{|x|-1}.$$

2. $\dfrac{2\left(\sqrt{5}+\sqrt{x}\right)}{5-x}$ Multiply numerator and denominator by the conjugate of the denominator:

$$\dfrac{2}{\sqrt{5}-\sqrt{x}}\cdot\dfrac{\sqrt{5}+\sqrt{x}}{\sqrt{5}+\sqrt{x}}=\dfrac{2\left(\sqrt{5}+\sqrt{x}\right)}{5-\sqrt{x^2}}=\dfrac{2\left(\sqrt{5}+\sqrt{x}\right)}{5-|x|}.$$

3. $x=7$ Square both sides of the equation to get $5x+1=11-x+2\cdot4\sqrt{11-x}+16$. Move the terms without a radical factor to the left to get $6x-26=8\sqrt{11-x}$. Divide each term by 2 before squaring both sides again: $(3x-13)^2=\left(4\sqrt{11-x}\right)^2$; $9x^2-78x+169=16(11-x)$. Distribute on the right; then move all the terms to the left: $9x^2-78x+169=176-16x$; $9x^2-62x-7=0$. Solve the quadratic by factoring or using the quadratic formula: $9x^2-62x-7=(9x+1)(x-7)=0$. Checking the two solutions, $x=-\dfrac{1}{9}$ and $x=7$, only the 7 works.

4. $x=\dfrac{1}{2}$ Square both sides of the equation to get $6x+1+2\sqrt{6x+1}\sqrt{2x+3}+2x+3=8x+12$. Move the terms without radicals to the right to get $2\sqrt{6x+1}\sqrt{2x+3}=8$. Divide each side by 2 before squaring both sides: $\left(\sqrt{6x+1}\sqrt{2x+3}\right)^2=(4)^2$; $(6x+1)(2x+3)=16$; $12x^2+20x+3=16$. Move the 16 to the left and then solve the quadratic: $12x^2+20x-13=(2x-1)(6x+13)=0$. The two solutions are $x=\dfrac{1}{2}$ and $x=-\dfrac{13}{6}$. Only the $x=\dfrac{1}{2}$ works.

5. $5\left(\sqrt{3}+\sqrt{2}\right)$ Subtract $x\sqrt{2}$ from each side to get the two terms with the variable in them on the same side. Then factor out the x: $5=x\sqrt{3}-x\sqrt{2}=x\left(\sqrt{3}-\sqrt{2}\right)$. Divide each side by the multiplier of x to get $x=\dfrac{5}{\sqrt{3}-\sqrt{2}}$. Rationalize the answer by multiplying numerator and denominator by the conjugate of the denominator: $x=\dfrac{5}{\sqrt{3}-\sqrt{2}}\cdot\dfrac{\sqrt{3}+\sqrt{2}}{\sqrt{3}+\sqrt{2}}=\dfrac{5\left(\sqrt{3}+\sqrt{2}\right)}{3-2}.$

Complex Numbers

Real numbers consist of *rational* numbers—those that can be written as a fraction and have a repeating or terminating decimal—and *irrational* numbers—those whose decimal value never repeats or terminates. Another type of number, a *complex* number, becomes necessary when there is no way to represent $\sqrt{-k}$, where the value under the radical was negative. $x^2-1=0$ has two solutions, ±1..., and we are pleased. $x^2+1=0$, however, has no real solutions, yet it is a perfectly nice equation. This is unpleasant. We name i, necessarily imaginary, a solution of this equation. $i=\sqrt{-1}$. The symbol i is used to denote $\sqrt{-1}$. The i stands for an *imaginary* number. So now $\sqrt{-4}$ can be written as $2i$, because $\sqrt{-4}=\sqrt{-1}\sqrt{4}=i\cdot2$. The general form for a complex number is $a+bi$. The letters a and b represent real numbers and $i=\sqrt{-1}$.

Example Problems

These problems show the answers and solutions.

1. Write the following as complex numbers: $\sqrt{-16}$ and $\sqrt{-7}$.

    answer: $4i$ and $\sqrt{7}\,i$

    The values under the radical can be written as a product of two radicals, and the two factors can be evaluated separately, $\sqrt{-16} = \sqrt{-1}\sqrt{16} = i \cdot 4$. The number 7 isn't a perfect square, so it's left under the radical: $\sqrt{-7} = \sqrt{-1}\sqrt{7} = i\sqrt{7}$.

2. Write the following as complex numbers: $\sqrt{-32}$ and $\dfrac{2 + \sqrt{4 - 4(25)}}{2}$.

    answer: $4\sqrt{2}\,i$ and $1 + 2\sqrt{6}\,i$

    The first radical has a factor that's a perfect square: $\sqrt{-32} = \sqrt{-1}\sqrt{16}\sqrt{2} = i \cdot 4 \cdot \sqrt{2}$. The second term has an expression under the radical that has to be simplified first.

    $$\frac{2 + \sqrt{4 - 4(25)}}{2} = \frac{2 + \sqrt{4 - 100}}{2} = \frac{2 + \sqrt{-96}}{2} = \frac{2 + \sqrt{-1}\sqrt{16}\sqrt{6}}{2} = \frac{\cancel{2}^{1} + i \cdot \cancel{4}^{2}\sqrt{6}}{\cancel{2}^{1}}$$

Work Problems

Use these problems to give yourself more practice.

1. Write $\sqrt{-9}$ as a complex number.

2. Write $\sqrt{-125}$ as a complex number.

3. Write $\sqrt{9 - 5(18)}$ as a complex number.

4. Write $\dfrac{4 \pm \sqrt{16 - 4(5)}}{2}$ as a complex number.

5. Write $\dfrac{-5 \pm \sqrt{25 - 4(13)}}{2}$ as a complex number.

Worked Solutions

1. **$3i$** $\sqrt{-9} = \sqrt{-1}\sqrt{9} = i \cdot 3$

2. **$5\sqrt{5}\,i$** $\sqrt{-125} = \sqrt{-1}\sqrt{25}\sqrt{5} = i \cdot 5 \cdot \sqrt{5}$

3. **$9i$** $\sqrt{9 - 5(18)} = \sqrt{9 - 90} = \sqrt{-81} = \sqrt{-1}\sqrt{81} = i \cdot 9$

4. **$2 \pm i$** $\dfrac{4 \pm \sqrt{16 - 4(5)}}{2} = \dfrac{4 \pm \sqrt{16 - 20}}{2} = \dfrac{4 \pm \sqrt{-4}}{2} = \dfrac{4 \pm 2i}{2} = 2 \pm i$

5. **$\dfrac{-5}{2} \pm \dfrac{3\sqrt{3}}{2}\,i$** $\dfrac{-5 \pm \sqrt{25 - 4(13)}}{2} = \dfrac{-5 \pm \sqrt{25 - 52}}{2} = \dfrac{-5 \pm \sqrt{-27}}{2} =$

    $\dfrac{-5 \pm \sqrt{-1}\sqrt{9}\sqrt{3}}{2} = \dfrac{-5 \pm 3i\sqrt{3}}{2}$

Operations Involving Complex Numbers

When complex numbers are in their $a + bi$ standard form, they can be combined using the familiar algebraic rules used in the operations of addition, subtraction, and multiplication. Division can be done very simply when a real number divides a complex number, but dividing two complex numbers is actually accomplished using multiplication. The numerator and denominator are multiplied by the conjugate of the denominator. There are some interesting twists that occur in the multiplication of two complex numbers. These occur because of what happens to powers of i. No rules for combining expressions with a negative under the radical exist, so these expressions have to be changed to the standard complex form first. Listed here are some powers of i and the results of operations on complex numbers. Note that the final result, in each case, is in the standard $a + bi$ form, where a and b are real numbers.

Powers of i: $i^1 = i = 0 + 1i$

$$i^2 = \left(\sqrt{-1}\right)^2 = -1 = -1 + 0i$$

$$i^3 = i^1 \cdot i^2 = i(-1) = -i = 0 - 1i$$

$$i^4 = i^2 \cdot i^2 = (-1)(-1) = 1 = 1 + 0i$$

Addition: $(a + bi) + (c + di) = (a + c) + (b + d)i$

Subtraction: $(a + bi) - (c + di) = (a - c) + (b - d)i$

Multiplication: $(a + bi)(c + di) = ac + adi + bci + bdi^2 = ac + (ad + bc)i + bd(-1) =$
$(ac - bd) + (ad + bc)i$

Division: $(a + bi) \div (c + di) = \dfrac{a + bi}{c + di} \cdot \dfrac{c - di}{c - di} = \dfrac{ac - adi + bci - bdi^2}{c^2 - d^2 i^2}$

$$= \dfrac{ac + (bc - ad)i - bd(-1)}{c^2 - d^2 i^2}$$

$$= \dfrac{(ac + bd) + (bc - ad)i}{c^2 + d^2} = \dfrac{ac + bd}{c^2 + d^2} + \dfrac{bc - ad}{c^2 + d^2}i$$

In the multiplication and division rules, the factor i^2 has been replaced by -1 so there are just real terms and terms with a factor of i.

The following examples show you how these work with numerical values. Rather than memorize the rules given here, you should just perform the operations as usual and simplify the answers to the standard complex number form.

Example Problems

These problems show the answers and solutions.

1. Add $4 + 5i$ and $6 - 3i$.

    answer: $10 + 2i$

    Rearrange the terms and combine like terms: $(4 + 5i) + (6 - 3i) = 4 + 6 + 5i - 3i = 10 + 2i$.

2. Subtract $4 + 5i$ and $6 - 3i$.

    answer: $-2 + 8i$

    Distribute the negative sign through the second number. Then combine like terms:
    $(4 + 5i) - (6 - 3i) = 4 + 5i - 6 + 3i = 4 - 6 + 5i + 3i = -2 + 8i$.

3.   Multiply $4 + 5i$ and $6 - 3i$.

     answer: $39 + 18i$

     Multiply the two expressions using FOIL to get $(4 + 5i)(6 - 3i) = 24 - 12i + 30i - 15i^2 = 24 - 15(-1) + (-12 + 30)i = 39 + 18i$.

4.   Divide $4 + 5i$ and $6 - 3i$.

     answer: $\dfrac{1}{5} + \dfrac{14}{15}i$

     Multiply numerator and denominator by the conjugate of the denominator: $\dfrac{4 + 5i}{6 - 3i} \cdot \dfrac{6 + 3i}{6 + 3i} = \dfrac{24 + 12i + 30i + 15i^2}{36 - 9i^2} = \dfrac{24 - 15 + (12 + 30)i}{36 + 9} = \dfrac{9 + 42i}{45} = \dfrac{9}{45} + \dfrac{42}{45}i = \dfrac{1}{5} + \dfrac{14}{15}i$. What if you used the "rule" for division, given previously?

     $$\frac{a + bi}{c + di} \cdot \frac{c - di}{c - di} = \frac{ac + bd}{c^2 + d^2} + \frac{bc - ad}{c^2 + d^2}i$$

     In this problem, $a = 4$, $b = 5$, $c = 6$, $d = -3$. Filling in the blanks, $\dfrac{4 + 5i}{6 - 3i} \cdot \dfrac{6 + 3i}{6 + 3i} = \dfrac{4 \cdot 6 + 5(-3)}{6^2 + (-3)^2} + \dfrac{5 \cdot 6 - 4(-3)}{6^2 + (-3)^2}i = \dfrac{24 - 15}{36 + 9} + \dfrac{30 + 12}{36 + 9}i = \dfrac{9}{45} + \dfrac{42}{45}i$. It's still easier just to do the multiplication.

5.   Add, subtract, multiply, and divide $\sqrt{-16}$ and $\sqrt{-4}$.

     answer: $6i$, $2i$, -8, 2

     In each case, the numbers have to be changed to their complex form before combining. $\sqrt{-16} = \sqrt{-1}\sqrt{16} = 4i$ and $\sqrt{-4} = \sqrt{-1}\sqrt{4} = 2i$. Add, $4i + 2i = 6i$. Subtract, $4i - 2i = 2i$. Multiply, $4i \cdot 2i = 8i^2 = 8(-1) = -8$. Divide, $\dfrac{4i}{2i} = 2$. You might be leery of this division problem—doing it without using a conjugate. Look at the same problem using the conjugate of the denominator: $\dfrac{4i}{2i} \cdot \dfrac{-2i}{-2i} = \dfrac{-8i^2}{-4i^2} = \dfrac{-8(-1)}{-4(-1)} = \dfrac{8}{4} = 2$.

6.   Find the value of i^{423}.

     answer: $-i$

     You do not have to do 423 multiplications to get this answer. There's a very nice pattern to the powers of i. The first four powers are shown at the beginning of this section. Here are some others for you: $i^1 = i$, $i^2 = -1$, $i^3 = -i$, $i^4 = 1$, $i^5 = i$, $i^6 = -1$, $i^7 = -i$, $i^8 = 1$, $i^9 = i$, $i^{10} = -1$, $i^{11} = -i$, $i^{12} = 1$, $i^{13} = i$, $i^{14} = -1$. If the power is a multiple of 4, then the term is equal to 1. One more than that power is i; two more than that is -1; and three more than that is $-i$. Since 423 is three more than 420, which is a multiple of 4, then $i^{423} = -i$.

Work Problems

Use these problems to give yourself additional practice.

1. Add $-6 + 8i$ and $4 - 5i$.

2. Subtract $-6 + 8i$ and $4 - 5i$.

3. Multiply $-6 + 8i$ and $4 - 5i$.

4. Divide $-6 + 8i$ and $4 - 5i$.

5. Add, subtract, multiply, and divide $\sqrt{-9}$ and $-\sqrt{-25}$.

Worked Solutions

1. **$-2 + 3i$** $(-6 + 8i) + (4 - 5i) = -6 + 4 + 8i - 5i = -2 + 3i$.

2. **$-10 + 13i$** $(-6 + 8i) - (4 - 5i) = -6 + 8i - 4 + 5i = -6 - 4 + 8i + 5i = -10 + 13i$

3. **$16 + 62i$** $(-6 + 8i)(4 - 5i) = -24 + 62i - 40i^2 = -24 + 40 + 62i = 16 + 62i$

4. $-\dfrac{64}{41} + \dfrac{2}{41}i$ $\dfrac{-6 + 8i}{4 - 5i} \cdot \dfrac{4 + 5i}{4 + 5i} = \dfrac{-24 - 30i + 32i + 40i^2}{16 - 25i^2} = \dfrac{-24 - 40 + 2i}{16 + 25} = \dfrac{-64 + 2i}{41}$

5. $-2i,\ 8i,\ 15,\ -\dfrac{3}{5}$ First change the two numbers to their complex form: $\sqrt{-9} = 3i$ and $-\sqrt{-25} = -5i$. Add, $3i + (-5i) = -2i$. Subtract, $3i - (-5i) = 3i + 5i = 8i$. Multiply, $3i(-5i) = -15i^2 = -15(-1) = 15$. Divide, $\dfrac{3i}{-5i} = -\dfrac{3}{5}$.

Quadratic Formula and Complex Numbers

The quadratic formula can be used to solve the quadratic equation $ax^2 + bx + c = 0$, and the solutions are $x = \dfrac{-b \pm \sqrt{b^2 - 4ac}}{2a}$. When using this formula, the value under the radical could turn out to be negative. Using $i = \sqrt{-1}$, the result can be written in the form of a complex number rather than leaving it incomplete. If there is a negative under the radical, then there's still no real solution, but a complex solution can also have meaning. Complex numbers are studied in advanced mathematics and have their own applications. For now, they'll be used to just finish the problem.

Example Problems

These problems show the answers and solutions.

1. Solve $x^2 - 2x + 5 = 0$.

    answer: $1 \pm 2i$

    Using the quadratic formula, $x = \dfrac{2 \pm \sqrt{(-2)^2 - 4(1)(5)}}{2(1)} = \dfrac{2 \pm \sqrt{4 - 20}}{2} = \dfrac{2 \pm \sqrt{-16}}{2} = \dfrac{2 \pm 4i}{2} = \dfrac{2}{2} \pm \dfrac{4i}{2}$.

2. Solve $3x^2 + 4x + 8 = 0$.

answer: $-\dfrac{2}{3} \pm \dfrac{2\sqrt{5}}{3}i$

Using the quadratic formula, $x = \dfrac{-4 \pm \sqrt{4^2 - 4(3)(8)}}{2(3)} = \dfrac{-4 \pm \sqrt{16 - 96}}{6} = \dfrac{-4 \pm \sqrt{-80}}{6}$

$= \dfrac{-4 \pm \sqrt{-1}\sqrt{16}\sqrt{5}}{6} = \dfrac{-4 \pm 4i\sqrt{5}}{6} = -\dfrac{4}{6} \pm \dfrac{4\sqrt{5}}{6}i$.

Work Problems

Use these problems to give yourself additional practice.

1. Solve $x^2 + 2x + 11 = 0$.

2. Solve $x^2 - 3x + 9 = 0$.

3. Solve $x^2 + 9 = 0$.

4. Solve $2x^2 - 5x + 11 = 0$.

5. Solve $3x^2 + 8x + 8 = 0$.

Worked Solutions

1. $-1 \pm \sqrt{10}\,i$ Using the quadratic formula, $x = \dfrac{-2 \pm \sqrt{2^3 - 4(1)(11)}}{2(1)} = \dfrac{-2 \pm \sqrt{4 - 44}}{2}$

$= \dfrac{-2 \pm \sqrt{-40}}{2} = \dfrac{-2 \pm \sqrt{-1}\sqrt{4}\sqrt{10}}{2} = \dfrac{-2 \pm 2i\sqrt{10}}{2} = -\dfrac{2}{2} \pm \dfrac{2\sqrt{10}}{2}i$.

2. $\dfrac{3}{2} \pm \dfrac{3\sqrt{3}}{2}i$ Using the quadratic formula, $x = \dfrac{3 \pm \sqrt{3^2 - 4(1)(9)}}{2(1)} = \dfrac{3 \pm \sqrt{9 - 36}}{2}$

$= \dfrac{3 \pm \sqrt{-27}}{2} = \dfrac{3 \pm \sqrt{-1}\sqrt{9}\sqrt{3}}{2} = \dfrac{3 \pm 3i\sqrt{3}}{2} = \dfrac{3}{2} \pm \dfrac{3\sqrt{3}}{2}i$.

3. $\pm 3i$ Using the quadratic formula, $x = \dfrac{0 \pm \sqrt{0^2 - 4(1)(9)}}{2(1)} = \dfrac{0 \pm \sqrt{-36}}{2} = \pm\dfrac{6i}{2}$.

4. $\dfrac{5}{4} \pm \dfrac{3\sqrt{7}}{4}i$ Using the quadratic formula, $x = \dfrac{5 \pm \sqrt{5^2 - 4(2)(11)}}{2(2)} = \dfrac{5 \pm \sqrt{25 - 88}}{4}$

$= \dfrac{5 \pm \sqrt{-63}}{4} = \dfrac{5 \pm \sqrt{-1}\sqrt{9}\sqrt{7}}{4} = \dfrac{5 \pm 3i\sqrt{7}}{4} = \dfrac{5}{4} \pm \dfrac{3\sqrt{7}}{4}i$.

5. $-\dfrac{4}{3} \pm \dfrac{2\sqrt{2}}{3}i$ Using the quadratic formula, $x = \dfrac{-8 \pm \sqrt{8^2 - 4(3)(8)}}{2(3)} = \dfrac{-8 \pm \sqrt{64 - 96}}{6}$

$= \dfrac{-8 \pm \sqrt{-32}}{6} = \dfrac{-8 \pm \sqrt{-1}\sqrt{16}\sqrt{2}}{6} = \dfrac{-8 \pm 4i\sqrt{2}}{6} = -\dfrac{8}{6} \pm \dfrac{4\sqrt{2}}{6}i$.

Rational Equations

A rational equation is one that contains one or more fractions in which the variable is in the denominator. Special care needs to be taken when solving these equations. The most convenient method for solving them involves getting rid of the fractions—multiplying through by a common denominator. This makes the equation much nicer to deal with, but it can result in introducing extraneous solutions. These extraneous solutions occur in other solutions, too. You'll see them when solving radical equations (and squaring both sides), exponential equations (when extracting the exponents), and logarithmic equations (when changing to exponential). Even with the complications that arise from changing the form of an equation, it's still usually the easiest and most efficient method of solving them. You just have to be aware of the possible problem and check the answer for those extraneous solutions.

Example Problems

These problems show the answers and solutions.

1. Solve $\frac{4}{x} - \frac{3}{x+2} = \frac{2x+1}{2x}$.

   answer: $x = 2$, $x = -\frac{7}{2}$

   The common denominator for the three fractions is $2x(x + 2)$. Multiplying each term by that product will eliminate the fractions. The extraneous solutions to watch out for are $x = 0$ and $x = -2$. Either one will result in a denominator equaling 0.

   $$\frac{4}{\cancel{x}} \cdot \frac{\cancel{2}x(x+2)}{1} - \frac{3}{\cancel{x+2}} \cdot \frac{2x\cancel{(x+2)}}{1} = \frac{2x+1}{\cancel{2x}} \cdot \frac{\cancel{2}x(x+2)}{1},$$
   $$4 \cdot 2(x+2) - 3 \cdot 2x = (2x+1)(x+2)$$

   Now distribute, multiply, and simplify the equation to get a quadratic equation to be solved: $8x + 16 - 6x = 2x^2 + 5x + 2$; $2x^2 + 3x - 14 = 0$. This equation can be solved by factoring or by using the quadratic formula. Factor it, $2x^2 + 3x - 14 = (x - 2)(2x + 7) = 0$. Set the factors equal to 0, $x = 2$ or $x = -\frac{7}{2}$. Both of these work. Neither causes the original denominators to be equal to 0.

2. Solve $\frac{2x-3}{x-1} = \frac{x+5}{x-1} - \frac{8}{(x-1)(x-2)}$.

   answer: $x = 4$, $x = 6$

   The common denominator for the three fractions is $(x - 1)(x - 2)$. Multiply each term by that common denominator:

   $$\frac{2x-3}{\cancel{x-1}} \cdot \frac{\cancel{(x-1)}(x-2)}{1} = \frac{x+5}{\cancel{x-1}} \cdot \frac{\cancel{(x-1)}(x-2)}{1} - \frac{8}{\cancel{(x-1)(x-2)}} \cdot \frac{\cancel{(x-1)(x-2)}}{1}$$

   Simplify the equation, $(2x - 3)(x - 2) = (x + 5)(x - 2) - 8$. Then multiply, simplify, and rewrite as a quadratic equation: $2x^2 - 7x + 6 = x^2 + 3x - 10 - 8$; $x^2 - 10x + 24 = 0$. This can be solved by factoring: $x^2 - 10x + 24 = (x - 4)(x - 6) = 0$. The two solutions are $x = 4$ and $x = 6$.

Work Problems

Use these problems to give yourself additional practice.

1. Solve $\dfrac{4}{x-2} + \dfrac{5}{x+1} = \dfrac{8}{x-1} + \dfrac{16x}{(x-1)(x-2)(x+1)}$

2. Solve $\dfrac{6}{x} - \dfrac{5}{x+1} = \dfrac{x-2}{4}$.

3. Solve $\dfrac{4}{x+2} - \dfrac{3}{2} = \dfrac{5}{x+3}$.

4. Solve $\dfrac{3}{x+1} + \dfrac{2x+1}{x} = \dfrac{5+2x}{x+1}$.

5. Solve $\dfrac{2x}{x+2} + \dfrac{x+6}{(x+2)(x+3)} = 0$.

Worked Solutions

1. **$x = 22$** Multiply each term by the common denominator.

$$\frac{4}{x-2} \cdot \frac{(x-1)(x-2)(x+1)}{1} + \frac{5}{x+1} \cdot \frac{(x-1)(x-2)(x+1)}{1}$$

$$= \frac{8}{x-1} \cdot \frac{(x-1)(x-2)(x+1)}{1} + \frac{16x}{(x-1)(x-2)(x+1)} \cdot \frac{(x-1)(x-2)(x+1)}{1}$$

Rewrite the new equation: $4(x-1)(x+1) + 5(x-1)(x-2) = 8(x-2)(x+1) + 16x$. Now multiply out the terms and combine like terms on one side of the equation: $4(x^2-1) + 5(x^2-3x+2) = 8(x^2-x-2) + 16x$; $4x^2 - 4 + 5x^2 - 15x + 10 = 8x^2 - 8x - 16 + 16x$; $x^2 - 23x + 22 = 0$. Factor the quadratic to get the solutions, $x^2 - 23x + 22 = (x-1)(x-22) = 0$. The solutions of the quadratic are $x = 1$ and $x = 22$, but $x = 1$ can't be used, because it creates a 0 in the denominator of the second fraction in the original problem.

2. **$x = 4$** Multiply each term by the common denominator.

$$\frac{6}{x} \cdot \frac{4x(x+1)}{1} - \frac{5}{x+1} \cdot \frac{4x(x+1)}{1} = \frac{x-2}{4} \cdot \frac{4x(x+1)}{1}$$

Now multiply out the terms and simplify them: $24(x+1) - 20x = (x-2)x(x+1)$; $24x + 24 - 20x = x^3 - x^2 - 2x$; $x^3 - x^2 - 6x - 24 = 0$. Using the Rational Root Theorem, the only real solution of this equation is $x = 4$, and it works in the original equation.

3. **$x = -\dfrac{14}{3}$, $x = -1$** Multiply each term by the common denominator:

$$\frac{4}{x+2} \cdot \frac{2(x+2)(x+3)}{1} - \frac{3}{2} \cdot \frac{2(x+2)(x+3)}{1} = \frac{5}{x+3} \cdot \frac{2(x+2)(x+3)}{1}$$

Multiply and simplify: $8(x+3) - 3(x+2)(x+3) = 10(x+2)$, $8x + 24 - 3(x^2 + 5x + 6) = 10x + 20$; $8x + 24 - 3x^2 - 15x - 18 = 10x + 20$; $3x^2 + 17x + 14 = 0$. Factor the quadratic to obtain the solutions, $3x^2 + 17x + 14 = (3x + 14)(x + 1) = 0$. The solutions are $x = -\dfrac{14}{3}$ and $x = -1$.

4. **No solution**　Multiply through by the common denominator:

$$\frac{3}{\cancel{x+1}} \cdot \frac{x(\cancel{x+1})}{1} + \frac{2x+1}{\cancel{x}} \cdot \frac{\cancel{x}(x+1)}{1} = \frac{5+2x}{\cancel{x+1}} \cdot \frac{x(\cancel{x+1})}{1}$$

Now multiply and simplify: $3x + (2x + 1)(x + 1) = (5 + 2x)x$, $3x + 2x^2 + 3x + 1 = 5x + 2x^2$; $6x + 1 = 5x$, $-x = 1$ or $x = -1$. This is a solution of the equation obtained by getting rid of the fractions, but it isn't a solution for the original equation. If $x = -1$, there would be a 0 in the denominator of two of the fractions.

5. $x = -\dfrac{3}{2}$　Multiply through by the common denominator:

$$\frac{2x}{\cancel{x+2}} \cdot \frac{(\cancel{x+2})(x+3)}{1} + \frac{x+6}{(x+2)\cancel{(x+3)}} \cdot \frac{(x+2)\cancel{(x+3)}}{1} = 0$$

Now multiply and simplify: $2x(x + 3) + x + 6 = 2x^2 + 6x + x + 6 = 2x^2 + 7x + 6 = 0$. Factor the quadratic to get $2x^2 + 7x + 6 = (2x + 3)(x + 2) = 0$. The solutions are $x = -\dfrac{3}{2}$ and $x = -2$. Only the first solution will work, because if $x = -2$, then the fractions would have 0 in their denominators.

Variation

A *proportion* is an equation in which two fractions or ratios are set equal to one another. $\dfrac{x}{3} = \dfrac{8}{6}$ is a proportion, and x can have only one value for it to be a true statement: $x = 4$. One property of proportions is the *cross-product* property. If $\dfrac{a}{b} = \dfrac{c}{d}$, then $ad = bc$. This is very useful when solving applications involving proportions and ratios.

An interesting result of proportions is *variations*. In variations, two variables can vary *directly* or *inversely* with one another. For example, the circumference of a circle varies directly with the diameter: $C = \pi d$. As the diameter gets bigger, so does the circumference. The perimeter of a square varies directly with the length of a side: $P = 4s$. As the side gets bigger, so does the perimeter. In these two cases, the π and the 4 are constants; they're called the *constant of proportionality* for that particular situation.

In general, direct variation means that if "y varies directly with x", then $y = kx$ where k is the constant of proportionality. With the other type of variation, inverse variation, if "y varies inversely with x," then $y = \dfrac{k}{x}$. In this case, as x gets larger, y gets smaller. An example of inverse variation would be in the case of the loudness of music. The distance from the music and the loudness vary inversely. The greater the distance, the less loudness there is.

Example Problems

These problems show the answers and solutions.

1. The distance you travel in a *given* amount of time varies directly with the speed you're traveling during that time. If you travel 120 miles when you're moving at 40 mph, then how far will you go if you're moving at 60 mph?

   answer: 180 miles

The formula for distance is $d = rt$. In this case, the amount of time is set—it's constant, so the formula becomes $d = kr$ where k is the constant of proportionality. You travel 120 miles at 40 mph, so $120 = k \cdot 40$. Solving this, $k = 3$. Now replace the k with 3 in the formula to get $d = 3r$. The question asks how far you'll go at 60 mph, so replace the r with 60 to get $d = 3 \cdot 60 = 180$. The faster you travel, the farther you go—that's direct variation.

2. The light intensity from a lamp varies inversely with the square of the distance from the light source. If the intensity from a lamp is 80 lumens at a distance of 20 feet, then what is the intensity at 40 feet?

    answer: 20 lumens

    The formula for intensity is $I = \dfrac{k}{d^2}$. Replacing the I with 80 and the d with 20, $80 = \dfrac{k}{20^2} = \dfrac{k}{400}$. Solving for k, $k = 32,000$. So the formula for this particular situation is $I = \dfrac{32,000}{d^2}$. To find the intensity at 40 feet, replace the d with 40 to get $I = \dfrac{32,000}{40^2} = \dfrac{32,000}{1,600} = 20$. The intensity decreased as the distance became greater.

3. The volume of a gas varies directly with the temperature and inversely with the pressure. When the temperature of a particular gas is 300°F, the pressure is 45 pounds per square inch and the volume is 15 cubic feet. What is the volume when the pressure decreases to 30 pounds per square inch, and the temperature increases to 330°F?

    answer: 24.75 cubic feet

    The formula for the volume is $V = \dfrac{kt}{p}$. The volume varies directly with the temperature, so t is in the numerator of the fraction. It varies inversely with the pressure, so p is in the denominator. The constant of proportionality is k. The first thing to do is to solve for k for this particular gas, $15 = \dfrac{k(300)}{45}$. Reduce the fraction. Then multiply each side by the reciprocal of the coefficient of k: $15 = \dfrac{k\left(\cancel{300}^{20}\right)}{\cancel{45}_{3}}$, $\dfrac{3}{\cancel{20}_{4}} \cdot \cancel{15}^{3} = \dfrac{\cancel{20}k}{\cancel{3}} \cdot \dfrac{\cancel{3}}{\cancel{20}}$, so $k = \dfrac{9}{4} = 2.25$.

    Now write the formula for this gas as $V = \dfrac{2.25t}{p}$. Replace the t with 330 and the p with 30 to get $V = \dfrac{2.25(330)}{30} = 2.25(11) = 24.75$.

Work Problems

Use these problems to give yourself additional practice.

1. y varies directly with x. If $x = 50$, then $y = 30$. What is y if x is 60?

2. y varies inversely with x. If $x = 20$, then $y = 80$. What is y if x is 40?

3. The distance an object will fall in t seconds varies directly with t^2. If an object falls 32 feet in one second, how long will it take for it to fall 128 feet?

4. The number of hours it takes to clean the dormitory varies inversely with the number of people cleaning it and directly with the number of people living there. If it takes 8 hours for 5 people to clean the dormitory when there are 100 people there, then how long will it take if there are 10 people cleaning (the number of people living there stays the same)?

5. In the preceding problem, how long will it take the original 5 people to clean the dormitory if there are now 120 people living in the dormitory?

Worked Solutions

1. **36**   Let $y = kx$. Replacing y and x to solve for k, $30 = k \cdot 50$, $k = \frac{30}{50} = \frac{3}{5}$. Now write the equation as $y = \frac{3}{5}x$. When $x = 60$, then $y = \frac{3}{5} \cdot 60 = 36$.

2. **40**   Let $y = \frac{k}{x}$. Replacing y and x to solve for k, $80 = \frac{k}{20}$, $k = 1,600$. Now write the equation as $y = \frac{1600}{x}$. When $x = 40$, then $y = \frac{1,600}{40} = 40$.

3. **2 seconds**   The formula reads $d = kt^2$. Replacing d and t to solve for k, $32 = k(1^2)$, $k = 32$. Now write the equation as $d = 32t^2$. When $d = 128$, $128 = 32t^2$. Divide each side by 32, and then take the square root of each side: $\frac{128}{32} = t^2$, $t^2 = 4$, $t = 2$. There are actually two solutions to that equation, but an answer of -2 seconds doesn't make sense here.

4. **4 hours**   The formula needed is $H = \frac{kp}{c}$ where H is the number of hours, p is the number of people living in the dormitory, and c is the number of people cleaning. Solving for k, $8 = \frac{k(100)}{5}$, $8 = 20k$, $k = \frac{8}{20} = \frac{2}{5} = 0.4$. Now the formula reads $H = \frac{0.4p}{c}$. If there are 10 people cleaning and 100 people living in the dormitory, $H = \frac{0.4(100)}{10} = 0.4(10) = 4$.

5. **9.6 hours**   Using the same formula, $H = \frac{0.4p}{c}$, now there are 120 people in the dormitory and 5 people cleaning, so $H = \frac{0.4(120)}{5} = 0.4(24) = 9.6$.

Chapter 8
Conic Sections

Conic sections have been of great interest to mathematicians for centuries. The name *conic sections* for these curves comes from the figures that are observed when cones are sliced by planes at different angles. The four conic sections are: circle, ellipse, parabola, and hyperbola.

Circle

A circle is a collection of points that are all the same distance from a particular point, called the *center*. A segment drawn from the center of a circle to any point on the circle is called the *radius*. The length of this radius is used to compute the area and circumference of the circle. A segment drawn from one point on the circle to another point on the circle, which also goes through the center, is call the *diameter*. The length of the diameter is twice that of the radius.

Area and Circumference

The area of a circle is found with the formula $A = \pi r^2$. The circumference (length around the outside) is $C = 2\pi r = \pi d$. The r represents the radius; the d represents the diameter; and π represents a number that is approximately 3.14159. Usually, either 3.14 or $\frac{22}{7}$ is used to approximate the value of π. In this section, the value 3.14 will be used for the examples and work problems.

Standard Form

The standard form for the equation of a circle on the coordinate plane is $(x - h)^2 + (y - k)^2 = r^2$ where h and k are the coordinates of the center, (h, k), of the circle and r is, of course, the radius. A circle whose equation is $(x - 2)^2 + (y + 7)^2 = 49$ can be written as $(x - 2)^2 + (y - (-7))^2 = 7^2$ so you can easily see that it has its center at $(2, -7)$ and has a radius of 7.

Example Problems

These problems show the answers and solutions.

1. Find the area and circumference of a circle with a radius of 5 inches.

   answer: The area, $A = 25\pi$ square inches; the circumference, $C = 10\pi$ inches.

   The formula for the area of a circle is $A = \pi r^2$. Substituting 5 for the r, $A = \pi(5^2) = 25\pi \approx$ 78.5 square inches. The formula for the circumference of a circle is $C = 2\pi r$. Substituting 5 for r, $C = 2\pi \cdot 5 = 10\pi \approx 31.4$ inches.

2. Find the area of a circle that has a circumference of 43.96 cm.

   answer: $A = 49\pi$ square centimeters

   The formula for the circumference of a circle is $C = 2\pi r$. Substitute the 43.96 in for C, replace the π with 3.14, and solve for r: $43.96 = 2(3.14)r$, $r = \dfrac{43.96}{2(3.14)} = \dfrac{43.96}{6.28} = 7$. The radius is 7 cm, so $A = \pi r^2 = \pi(7^2) = 49\pi \approx 153.86$ square centimeters.

3. What are the center and radius of a circle whose equation is $(x - 3)^2 + (y - 4)^2 = 25$?

   answer: The center is at (3,4), and the radius is 5.

   The standard form for the equation of a circle is $(x - h)^2 + (y - k)^2 = r^2$. The $h = 3$, $k = 4$, and $r = 5$.

4. What are the center and radius of a circle whose equation is $(x + 8)^2 + (y - 2)^2 = \dfrac{1}{4}$?

   answer: The center is at (−8,2) and the radius is $\dfrac{1}{2}$.

   This time, the standard form has a *double negative* in it. Think of the equation as being $\left(x - (-8)\right)^2 + (y - 2)^2 = \left(\dfrac{1}{2}\right)^2$. Now it's in the correct form, with subtractions in both parentheses. The center is at (−8,2). The square root of $\dfrac{1}{4}$ is $\dfrac{1}{2}$, so that's the radius.

Changing to the Standard Form Using Completing the Square

Sometimes, you're given the equation of a circle, but it isn't in the standard form. It's really difficult to tell anything about the circle unless it's in the standard form. If you're told that the equation of a circle is $x^2 + y^2 - 2x + 10y = 38$, can you determine the center and radius? They're not really apparent. This equation can be changed to the standard form using *completing the square*. If you need to review this process, see "Completing the Square" in Chapter 2.

Example Problems

These problems show the answers and solutions.

1. Use completing the square to change $x^2 + y^2 - 2x + 10y = 38$ to the standard form for a circle. Then determine the center and radius of the circle.

   answer: $(x - 1)^2 + (y + 5)^2 = 64$; center is at (1,−5); radius is 8.

   First rearrange the terms on the left so the x's and y's are together and so that a space is after each grouping to put a number: $x^2 - 2x + \quad y^2 + 10y \quad = 38$. Completing the square on $x^2 - 2x$, add a 1 to the grouping. And, to keep the equation balanced, add 1 to the right side: $x^2 - 2x + 1 + y^2 + 10y \quad = 38 + 1$. To complete the square on the $y^2 + 10y$, add 25 to that grouping and 25 to the right side: $x^2 - 2x + 1 + y^2 + 10y + 25 = 38 + 1 + 25$. Now the first three terms and the last three terms on the left can be factored as squares of binomials. For a review of factoring trinomials, see "Factoring Trinomials" in Chapter 2, and for more on perfect square trinomials, see "Special Products" in Chapter 1. Factoring on the left and simplifying on the right, you get $(x-1)^2 + (y+5)^2 = 64$. This is in the standard form. The center is at $(1,-5)$, and the 64 is the square of 8, so the radius is 8.

2. Use completing the square to change $4x^2 + 4y^2 + 16x - 24y = 29$ to the standard form for a circle. Then determine the center and radius.

answer: $(x+2)^2 + (y-3)^2 = \frac{81}{4}$; center is at $(-2,3)$; radius is $\frac{9}{2}$

First, rearrange the terms on the left so the x's and y's are together and so that a space is after each grouping to put a number: $4x^2 + 16x \quad + 4y^2 - 24y \quad = 29$. Before attempting to complete the square, first factor the two groupings by factoring out just the coefficient of the squared term: $4(x^2 + 4x \quad) + 4(y^2 - 6y \quad) = 29$. Complete the square in each case. For the x's, $4(x^2 + 4x \quad)$ becomes $4(x^2 + 4x + 4)$. Also, add 16 to the other side of the equation. You add 16, not 4, because that 4 added inside the parentheses is multiplied by the 4 outside the parentheses. For the y's, $4(y^2 - 6y \quad)$ becomes $4(y^2 - 6y + 9)$, and 36 is added to the other side of the equation. $4(x^2 + 4x + 4) + 4(y^2 - 6y + 9) = 29 + 16 + 36$ becomes $4(x+2)^2 + 4(y-3)^2 = 81$. Now divide each term by 4 to get $(x+2)^2 + (y-3)^2 = \frac{81}{4}$. The center is at $(-2,3)$, and the radius is $\frac{9}{2}$, because $\left(\frac{9}{2}\right)^2 = \frac{81}{4}$.

Work Problems

Use these problems to give yourself additional practice.

1. Find the diameter, area, and circumference of a circle with a radius of 10 yards.

2. Find the center, radius, and diameter of the circle $(x-3)^2 + y^2 = 100$.

3. Find the center, radius, area, and circumference of the circle $(x+2)^2 + (y+5)^2 = \frac{121}{25}$.

4. Find the standard form of the equation for the circle $x^2 + y^2 + 6x - 12y = 4$.

5. Find the standard form of the equation for the circle $9x^2 + 9y^2 - 18x - 72y = 16$.

Worked Solutions

1. **Diameter: 20 yards; area: 100π square yards; circumference: 20π yards** The diameter is twice the length of the radius, so $d = 2(10) = 20$ yards. The area is $A = \pi r^2 = \pi(10)^2 = 100\pi \approx 314$ square yards. The circumference is $C = 2\pi r = 2\pi(10) = 20\pi \approx 62.8$ yards.

2. **Center: $(3,0)$; radius: 10; diameter: 20** The second term is technically $(y-0)^2$, so the y-coordinate of the center is 0. The radius is 10, so the diameter is twice that or 20.

3. **Center: $(-2,-5)$; radius: $\frac{11}{5}$; area: $\frac{121}{25}\pi$ square units; circumference: $\frac{22}{5}\pi$ units** The center and radius are obtained from the standard form written as $(x-(-2))^2 + (y-(-5))^2 = \left(\frac{11}{5}\right)^2$. The area is $A = \pi r^2 = \pi\left(\frac{11}{5}\right)^2 = \frac{121}{25}\pi$. The circumference is $C = 2\pi r = 2\pi\left(\frac{11}{5}\right) = \frac{22}{5}\pi$.

4. **$(x + 3)^2 + (y − 6)^2 = 49$**   First group the x's and y's and leave a space to complete the squares of each grouping: $x^2 + 6x \quad + y^2 − 12y \quad = 4$. Complete the squares and add the same amount to each side of the equation: $x^2 + 6x + 9 + y^2 − 12y + 36 = 4 + 9 + 36$. Factor the left and simplify the right: $(x + 3)^2 + (y − 6)^2 = 49$. The center of the circle is at $(−3,6)$, and the radius is 7.

5. **$(x − 1)^2 + (y − 4)^2 = \dfrac{169}{9}$**   First group the x's and y's and leave a space to complete the squares of each grouping: $9x^2 − 18x \quad + 9y^2 − 72y \quad = 16$. Factor out the coefficient of the squared term in each grouping: $9\left(x^2 − 2x \quad \right) + 9\left(y^2 − 8y \quad \right) = 16$. Complete the squares and add the *products* to the right side: $9(x^2 − 2x + 1) + 9(y^2 − 8y + 16) = 16 + 9 + 144$. Factor on the left and simplify on the right to get $9(x − 1)^2 + 9(y − 4)^2 = 169$. Divide each term by 9: $(x − 1)^2 + (y − 4)^2 = \dfrac{169}{9}$. The center is at $(1,4)$, and the radius is $\dfrac{13}{3}$.

Ellipse

An ellipse or oval is a shape that is aesthetically pleasing, useful, and found in many applications. The paths of the planets are elliptical. An ellipse has a center like a circle, but the shape of the ellipse is determined by two points (foci) that are on either side of the center and the same distance from the center. Let the distance from a point on the ellipse to the first focus be d, and let the distance from that same point on the ellipse to the second focus be e. Then, for every point on the ellipse, $d + e$ is the same value. Individually, d and e will change, but they will add up to the same thing.

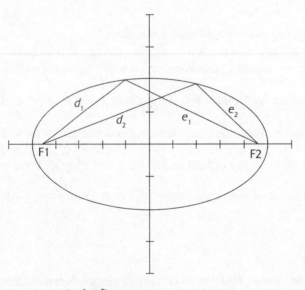

In the figure , $d_1 + e_1 = d_2 + e_2$

An ellipse has a horizontal axis and a vertical axis. They both go through the center and have endpoints on the ellipse. One axis is longer than the other. If the two axes had the same length, then you'd have a circle! The standard form of the equation for an ellipse is $\dfrac{(x − h)^2}{a^2} + \dfrac{(y − k)^2}{b^2} = 1$.

The length of the horizontal axis is $2a$, and the length of the vertical axis is $2b$. The h and k are the coordinates of the center, (h,k), of the ellipse, and the a and b are those values connected with the lengths of the axes. When the ellipse has its center at the origin, like the ellipse shown in the figure above, the foci are found at $(c,0)$ and $(−c,0)$ where $c^2 = a^2 − b^2$.

Sketching the graph of an ellipse usually involves finding the center, determining the endpoints of the axes, and sketching in an oval to fit those points.

Example Problems

These problems show the answers and solutions.

1. Find the center and lengths of the horizontal and vertical axes of the ellipse
$$\frac{(x-3)^2}{49} + \frac{(y+7)^2}{9} = 1.$$

   answer: Center at $(3, -7)$; horizontal axis: 14 units; vertical axis: 6 units

   The center is determined from the numerators of the fractions. The numerator of the second fraction could be written $(y-(-7))^2$ to fit the $(y-k)^2$ format. The value of a^2 is 49, so a is equal to 7, and $2a$ is equal to 14. The value of b^2 is 9, so b is equal to 3, and $2b$ is equal to 6.

2. Find the center and lengths of the horizontal and vertical axes of the ellipse $8(x+11)^2 + 32(y-1)^2 = 32$.

   answer: Center at $(-11, 1)$; horizontal axis: 4 units; vertical axis: 2 units

   The equation is not in standard form, so first divide each term by 32:
$$\frac{8(x+11)^2}{32} + \frac{32(y-1)^2}{32} = \frac{32}{32}; \frac{(x+11)^2}{4} + \frac{(y-1)^2}{1} = 1.$$ The center can be determined from the numerators of the fractions. The first fraction's numerator can be written $(x-(-11))^2$. The value of a^2 is 4, so a is 2 and $2a$ is 4. The value of b^2 is 1, so b is 1 and $2b$ is 2.

3. Use completing the square to write the equation $4x^2 + 25y^2 - 8x + 300y + 804 = 0$ in the standard form of an ellipse.

   answer: $\dfrac{(x-1)^2}{25} + \dfrac{(y+6)^2}{4} = 1$

   First, group the x terms together, the y terms together, and add -804 to each side to move the constant to the right: $4x^2 - 8x + 25y^2 + 300y = -804$. Factor 4, the coefficient of the x^2 term from the first two terms. Factor 25, the coefficient of the y^2 term, from the second two terms. Leave a space to complete the square in each case: $4(x^2 - 2x \quad) + 25(y^2 + 12y \quad) = -804$. Complete the square by adding 1 to $x^2 - 2x$ and $4 \cdot 1$ or 4 to the right side of the equation. Add 36 to $y^2 + 12y$ and $25 \cdot 36$ or 900 to the right side of the equation: $4(x^2 - 2x + 1) + 25(y^2 + 12y + 36) = -804 + 4 + 900$. Factor the two trinomials on the left and simplify on the right: $4(x-1)^2 + 25(y+6)^2 = 100$.

   Now divide each term by 100: $\dfrac{4(x-1)^2}{100} + \dfrac{25(y+6)^2}{100} = \dfrac{100}{100}$. Simplify the fractions, $\dfrac{(x-1)^2}{25} + \dfrac{(y+6)^2}{4} = 1$.

Work Problems

Use these problems to give yourself additional practice.

1. Find the center and the lengths of the two axes in $\dfrac{(x-2)^2}{25} + \dfrac{(y-2)^2}{64} = 1$.

2. Find the center and the lengths of the two axes in $9(x+3)^2 + 4(y+1)^2 = 36$.

3. Find the center and the lengths of the two axes in $16(x-7)^2 + 64y^2 = 1$.

4. Write the equation of the ellipse $x^2 + 100y^2 + 6x - 800y + 1509 = 0$ in standard form.

5. Write the equation of the ellipse $36x^2 + y^2 - 72x + 2y + 28 = 0$ in standard form.

Worked Solutions

1. **Center: (2,2); horizontal axis: 10 units; vertical axis: 16 units**   The coordinates of the center are obtained from the numerators of the fractions. The value of a^2 is 25, so a is equal to 5, and $2a$ is equal to 10. The value of b^2 is 64, so b is equal to 8, and $2b$ is equal to 16.

2. **Center: (−3,−1); horizontal axis: 4 units; vertical axis: 6 units**   First divide each term by 36: $\dfrac{9(x+3)^2}{36} + \dfrac{4(y+1)^2}{36} = \dfrac{36}{36}; \dfrac{(x+3)^2}{4} + \dfrac{(y+1)^2}{9} = 1$. The coordinates of the center are found from the numerators of the fractions which can be written as $(x - (-3))^2$ and $(y - (-1))^2$. The value of a^2 is 4, so a is equal to 2, and $2a$ is equal to 4. The value of b^2 is 9, so b is equal to 3, and $2b$ is equal to 6.

3. **Center: (7,0); horizontal axis: $\dfrac{1}{2}$ unit; vertical axis: $\dfrac{1}{4}$ unit**   First, write the equation in the standard form. The constant on the right is already 1, so all the work has to be done on the left. Rewrite the two terms on the left as fractions with the reciprocal of the coefficients in the denominators: $16(x-7)^2 + 64y^2 = \dfrac{(x-7)^2}{\text{\small$\frac{1}{16}$}} + \dfrac{y^2}{\text{\small$\frac{1}{64}$}} = 1$. This is a manipulation using complex fractions. See more on "Complex Fractions" in the Introduction. The center is determined from the numerators. The second fraction's numerator can be written as $(y - 0)^2$. The value of a^2 is $\dfrac{1}{16}$, so a is equal to $\dfrac{1}{4}$, and $2a$ is equal to $\dfrac{2}{4} = \dfrac{1}{2}$. The value of b^2 is $\dfrac{1}{64}$, so b is equal to $\dfrac{1}{8}$, and $2b$ is equal to $\dfrac{2}{8} = \dfrac{1}{4}$.

4. $\dfrac{(x+3)^2}{100} + \dfrac{(y-4)^2}{1} = 1$ First, group the x's together and the y's together and add -1509 to each side to move the constant to the right: $x^2 + 6x + 100y^2 - 800y = -1509$. Now factor 100 out of the last two terms and write each grouping with a space at the end to complete the square: $(x^2 + 6x \quad) + 100(y^2 - 8y \quad) = -1509$. Complete the square in each parentheses and add the amount needed to the right to keep the equation balanced: $(x^2 + 6x + 9) + 100(y^2 - 8y + 16) = -1509 + 9 + 1600$. Factor on the left and simplify on the right to get $(x + 3)^2 + 100(y - 4)^2 = 100$. Divide each term by 100: $\dfrac{(x+3)^2}{100} + \dfrac{100(y-4)^2}{100} = \dfrac{100}{100}; \dfrac{(x+3)^2}{100} + \dfrac{(y-4)^2}{1} = 1$.

5. $\dfrac{(x-1)^2}{\text{\small$\frac{1}{4}$}} + \dfrac{(y+1)^2}{9} = 1$ First, group the x's together, the y's together, and add -28 to each side to move the constant to the right: $36x^2 - 72x + y^2 + 2y = -28$. Factor out 36 from the first two terms. Then write the two groupings on the left as binomials with a space to complete the square and make the perfect square trinomials: $36(x^2 - 2x \quad) + (y^2 + 2y \quad) = -28$. Complete the square in each parentheses and add values on the right to keep the equation balanced: $36(x^2 - 2x + 1) + (y^2 + 2y + 1) = -28 + 36 + 1$. Factor the trinomials on the left and simplify the right side of the equation:

$36(x - 1)^2 + (y + 1)^2 = 9$. Divide each term by 9: $\dfrac{36(x-1)^2}{9} + \dfrac{(y+1)^2}{9} = \dfrac{9}{9}$. Simplifying, you get $\dfrac{4(x-1)^2}{1} + \dfrac{(y+1)^2}{9} = 1$. This is not in the standard form, because of the first fraction. Write the reciprocal of 4 in the denominator to complete the form. $\dfrac{(x-1)^2}{\frac{1}{4}} + \dfrac{(y+1)^2}{9} = 1$.

Parabola

The parabola is a U-shaped curve that can open upward, downward, to the right, or to the left. Those parabolas opening up or down are functions (see Chapter 3 on "Functions" for more on this), but the others are not. Parabolic shapes are found in dish antennas and reflectors of headlights. The standard forms for parabolas are as follows:

$y - k = a(x - h)^2$ These parabolas open upward if a is positive and downward if a is negative. The variable x has a squared term.

$x - h = a(y - k)^2$ These parabolas open to the right if a is positive and to the left if a is negative. The variable y has a squared term.

The value of a is also used to predict if the parabola is narrower than or wider/flatter than the basic parabola, $y = x^2$. When $|a|$ is greater than 1, then the parabola is narrower than $y = x^2$; the greater the value, the narrower the parabola. When $|a|$ is less than 1, then the parabola is wider or flatter than $y = x^2$.

The vertex of the parabola is (h,k).

Example Problems

These problems show the answers and solutions.

1. Determine the vertex and the opening direction of the parabola $y - 3 = 5(x + 4)^2$.

    answer: Vertex: $(-4,3)$; opens upward.

    The vertex is obtained from the general form where the number subtracted from x is the x coordinate, and the number subtracted from y is the y coordinate. The value in the parentheses can be written as $(x -(-4))^2$. Because the x value is squared, the parabola opens upward or downward. The positive 5 means it opens upward and is narrower than $y = x^2$.

2. Determine the vertex and opening direction of the parabola $x = -\frac{1}{2}(y + 2)^2 + 3$.

    answer: Vertex: $(3,-2)$; opens to the left.

    First subtract 3 from each side to put the equation in the general form: $x - 3 = -\frac{1}{2}(y + 2)^2$. The vertex is obtained from the number subtracted from the x and the number subtracted from the y. The value in the parentheses can be written as $(y -(-2))^2$. Because the y value is squared, the parabola opens right or left. The negative $\frac{1}{2}$ means that it opens left and is wider than $y = x^2$.

3. Use completing the square to find the standard form, vertex, and opening direction of the parabola $y = -2x^2 + 4x + 5$.

   answer: $y - 7 = -2(x - 1)^2$; vertex $(1,7)$; opens downward.

   First subtract 5 from each side and factor -2, the coefficient of the x^2 term, out of the two terms on the right: $y - 5 = -2(x^2 - 2x)$. Complete the square on the right and add -2 to the left to keep the equation balanced: $y - 5 - 2 = -2(x^2 - 2x + 1)$. Now simplify on the left and factor on the right to get $y - 7 = -2(x - 1)^2$. The x term is squared, so this is a parabola that opens upward or downward. Because the multiplier, -2, which corresponds to the a in the standard form, is negative, it opens downward and is narrower than $y = x^2$.

4. Use completing the square to find the standard form, vertex, and opening direction of the parabola $0 = y^2 - 2x - 6y + 5$.

   answer: $x + 2 = \frac{1}{2}(y - 3)^2$; vertex $(-2,3)$; opens to the right.

   Leave the two terms with y in them on the right and move the other two terms to the left by adding $2x$ to each side and subtracting 5 from each side: $2x - 5 = y^2 - 6y$. Complete the square on the right by adding 9; add 9 to the left: $2x - 5 + 9 = y^2 - 6y + 9$. Simplify on the left and factor on the right: $2x + 4 = (y - 3)^2$. Factor out a 2 from the terms on the left: $2(x + 2) = (y - 3)^2$. Divide each side by 2: $\frac{2(x + 2)}{2} = \frac{(y - 3)^2}{2}$. Simplify and write in the standard form: $x + 2 = \frac{1}{2}(y - 3)^2$. The vertex is $(-2,3)$. Because the y is squared, it opens left or right. The $\frac{1}{2}$ is positive, so it opens to the right and is wider/flatter than $y = x^2$.

Work Problems

Use these problems to give yourself additional practice.

1. Find the vertex and opening direction of the parabola $x + 2 = -\frac{1}{3}(y + 1)^2$.

2. Find the vertex and opening direction of the parabola $y = 4(x - 3)^2$.

3. Find the vertex and opening direction of the parabola $6 - 2y = 8(x + 4)^2$.

4. Change the equation $3y^2 + 36y + x + 113 = 0$ into the standard form for a parabola. Then find the vertex and opening direction.

5. Change the equation $x^2 + 8x - 6y + 40 = 0$ into the standard form for a parabola. Then find the vertex and opening direction.

Worked Solutions

1. **Vertex: $(-2,-1)$; opens to the left**   The coordinates of the vertex are the values subtracted from x and y, respectively. The equation could be written $x - (-2) = -\frac{1}{3}(y - (-1))^2$ to reflect the standard form. The y term is squared, so the parabola opens left or right. Because the $\frac{1}{3}$ is negative, it opens to the left and is wider/flatter than $y = x^2$.

2. **Vertex: (3,0); opens upward** The equation could be written $y - 0 = 4(x - 3)^2$ to better determine the vertex. The x term is squared, and the 4 is positive, so it opens upward.

3. **Vertex: (−4,3); opens downward** Factor out −2 from the terms on the left; then divide each side by −2 to get the standard form: $-2(y - 3) = 8(x + 4)^2$; $\dfrac{-2(y - 3)}{-2} = \dfrac{8(x + 4)^2}{-2}$; $y - 3 = -4(x + 4)^2$. The vertex is obtained from the numbers subtracted from the x and the y. The x is squared, and the 4 is negative, so the parabola opens downward and is narrower than $y = x^2$.

4. $x + 5 = -3(y + 6)^2$; **vertex: (−5, −6); opens to the left** Subtract $3y^2$ and $36y$ from each side to get them to the right side. Then factor out −3 from each of the terms on the right: $x + 113 = -3y^2 - 36y$; $x + 113 = -3(y^2 + 12y)$. Complete the square on the right. Add $-3 \cdot 36$ or −108 to the left side to keep the equation balanced: $x + 113 - 108 = -3(y^2 + 12y + 36)$. Simplify on the left and factor on the right: $x + 5 = -3(y + 6)^2$. The vertex is more apparent if you write the equation as $x - (-5) = -3(y - (-6))^2$. The y is squared, and the multiplier 3 is negative, so the parabola opens to the left and is narrower than $y = x^2$.

5. $y - 4 = \dfrac{1}{6}(x + 4)^2$; **vertex: (−4,4); opens upward** Add $6y$ and subtract 40 to get $x^2 + 8x = 6y - 40$. Then switch sides to work toward the standard form. You could subtract x^2 and $8x$ from each side, but then there are lots of negative signs. This is usually more efficient: $6y - 40 = x^2 + 8x$. Complete the square by adding 16 to each side: $6y - 40 + 16 = x^2 + 8x + 16$. Simplify on the left and factor on the right: $6y - 24 = (x + 4)^2$. Factor out 6 from the two terms on the left: $6(y - 4) = (x + 4)^2$. Then divide each side by 6: $\dfrac{6(y - 4)}{6} = \dfrac{(x + 4)^2}{6}$. This simplifies to the standard form $y - 4 = \dfrac{1}{6}(x + 4)^2$. The vertex is obtained from this form. The x is squared, and the $\dfrac{1}{6}$ is positive, so the parabola opens upward and is wider/flatter than $y = x^2$.

Hyperbola

The hyperbola looks something like two flattened cups sitting back-to-back. They can open left and right or up and down. The standard forms for a hyperbola are

$$\frac{(x - h)^2}{a^2} - \frac{(y - k)^2}{b^2} = 1 \quad \text{for hyperbolas opening left and right, and}$$

$$\frac{(y - k)^2}{b^2} - \frac{(x - h)^2}{a^2} = 1 \quad \text{for hyperbolas opening up and down.}$$

The (h,k) is the center, and $2a$ and $2b$ are the dimensions of the rectangle used to draw the hyperbola. The opening depends on whether the x terms or the y terms are negative. The use of these values is covered in the section in this chapter on "Graphing Conics."

Notice that the form for the hyperbola is different from that of an ellipse only by the operation sign; it's subtract instead of add, and the positive term determines how the hyperbola opens.

Example Problems

These problems show the answers and the solutions.

1. Find the center of the hyperbola $\dfrac{(x-3)^2}{4} - \dfrac{(y+3)^2}{9} = 1$ and determine which way it opens.

    answer: Center: $(3,-3)$; opens left and right.

    The center is obtained from the numerators of the fractions. The x^2 term is positive, so it opens left and right.

2. Find the center of the hyperbola $25y^2 - 16x^2 + 100y - 288x = 1596$ and determine which way it opens.

    answer: Center: $(-9,-2)$; opens upward and downward.

    First, the equation has to be written in standard form. Rearrange the terms so the y terms are grouped together and the x terms are grouped together. Then factor each grouping so the coefficient of the squared term is 1: $25y^2 + 100y - 16x^2 - 288x = 1596$; $25(y^2 + 4y) - 16(x^2 + 18x) = 1596$. Complete the square in each parentheses and add or subtract the amounts necessary to keep the equation balanced: $25(y^2 + 4y + 4) - 16(x^2 + 18x + 81) = 1596 + 100 - 1296$. Factor on the left and simplify on the right to get $25(y + 2)^2 - 16(x + 9)^2 = 400$. Divide each term by 400: $\dfrac{25(y+2)^2}{400} - \dfrac{16(x+9)^2}{400} = \dfrac{400}{400}$. Simplify to get the standard form, $\dfrac{(y+2)^2}{16} - \dfrac{(x+9)^2}{25} = 1$. The center is found by determining what is being subtracted from the x and the y. Another way to write the standard form is $\dfrac{(y-(-2))^2}{16} - \dfrac{(x-(-9))^2}{25} = 1$. The y^2 term is positive, so the hyperbola opens upward and downward.

Work Problems

Use these problems to give yourself additional practice.

1. Find the center of the hyperbola $\dfrac{\left(x-\frac{1}{4}\right)^2}{\frac{1}{4}} - \dfrac{\left(y-\frac{1}{9}\right)^2}{\frac{1}{9}} = 1$ and determine which way it opens.

2. Find the center of the hyperbola $(y-1)^2 - \dfrac{(x+2)^2}{5} = 1$ and determine which way it opens.

3. Find the center of the hyperbola $y^2 - 4x^2 = 4$ and determine which way it opens.

4. Use completing the square to write the hyperbola $x^2 - 81y^2 - 6x + 324y = 396$ in the standard form.

5. Use completing the square to write the hyperbola $4y^2 - x^2 + 8y + 18x = 79$ in the standard form.

Worked Solutions

1. **Center:** $\left(\frac{1}{4}, \frac{1}{9}\right)$**; opens left and right**   The center is determined by what is subtracted from the x and y in the standard form. The x term is positive, so the hyperbola opens left and right.

2. **Center: (−2,1); opens up and down**   The center is determined by what is subtracted from the x and y in the standard form. The y^2 term is negative, so the hyperbola opens up and down.

3. **Center: (0,0); opens up and down**   The standard form is found by dividing each term by 4. Then, the numerators of the fractions are rewritten as subtraction problems to emphasize the standard form for determining the center: $\dfrac{(y-0)^2}{4} - \dfrac{(x-0)^2}{1} = 1$. The y^2 term is negative, so the hyperbola opens up and down.

4. $\dfrac{(x-3)^2}{81} - \dfrac{(y-2)^2}{1} = 1$ Group the x's and y's together. Then factor out the coefficient of the squared term in each grouping: $(x^2 - 6x) - 81(y^2 - 4y) = 396$. Now complete the square in each parentheses and add or subtract the appropriate values to keep the equation balanced: $(x^2 - 6x + 9) - 81(y^2 - 4y + 4) = 396 + 9 - 324$. Factor on the left, and simplify on the right: $(x-3)^2 - 81(y-2)^2 = 81$. Divide each term by 81 and simplify: $\dfrac{(x-3)^2}{81} - \dfrac{81(y-2)^2}{81} = \dfrac{81}{81}; \dfrac{(x-3)^2}{81} - \dfrac{(y-2)^2}{1} = 1$

5. $\dfrac{(y+1)^2}{\frac{1}{2}} - \dfrac{(x-9)^2}{2} = 1$ Group the y's and x's together. Then factor out the coefficient of the squared term in each grouping: $4(y^2 + 2y) - 1(x^2 - 18x) = 79$. Complete the square in each parentheses and add or subtract the values necessary to keep the equation balanced: $4(y^2 + 2y + 1) - 1(x^2 - 18x + 81) = 79 + 4 - 81$. Factor on the left and simplify on the right: $4(y + 1)^2 - 1(x - 9)^2 = 2$. Divide each term by 2: $\dfrac{4(y+1)^2}{2} - \dfrac{1(x-9)^2}{2} = \dfrac{2}{2}$. Simplify each term. The first fraction will have the reciprocal of the multiplier 2 in the denominator to put the equation in the standard form: $\dfrac{2(y+1)^2}{1} - \dfrac{(x-9)^2}{2} = 1; \dfrac{(y+1)^2}{\frac{1}{2}} - \dfrac{(x-9)^2}{2} = 1$

Graphing Conics

Once the equation of a conic section is in its standard form, it's rather straight-forward and simple to recognize what it is and graph it. If you know the center and radius of a circle, you can sketch that circle in the correct position and make it the correct size. If you know the center and length of the axes of an ellipse or the center and a and b values of a hyperbola, then you can quickly sketch it. With the vertex, direction, and another point or two on a parabola, you can sketch it efficiently. One challenge is in recognizing which conic is which so that you can do that quick sketching. In their standard forms, the different conics are recognizable. Here are the standard forms of the conic sections:

Circle	$(x - h)^2 + (y - k)^2 = r^2$
Ellipse	$\dfrac{(x-h)^2}{a^2} + \dfrac{(y-k)^2}{b^2} = 1$
Parabola	$y - k = a(x - h)^2$ or $x - h = a(y - k)^2$
Hyperbola	$\dfrac{(x-h)^2}{a^2} - \dfrac{(y-k)^2}{b^2} = 1$ or $\dfrac{(y-k)^2}{b^2} - \dfrac{(x-h)^2}{a^2} = 1$

For details on these standard forms, refer to the respective sections in this chapter.

When the equation of a conic section isn't in its standard form, then you need to change it to that form using completing the square. You'll need to recognize which conic it is in order to work toward that form, so here are the rules for determining which conic is which. Consider the general conic equation $Ax^2 + By^2 + Cx + Dy + F = 0$ and just refer to the values of A and B.

If $A = B$, then you have a circle.

If $A \neq B$, and they have the same sign, then you have an ellipse.

If $A \neq B$, and they have different signs, then you have an hyperbola.

If $A = 0$ or $B = 0$ but not both (this is the same as saying that there's only one squared term), then you have a parabola.

Example Problems

These problems show the answers and solutions.

1. Sketch the graph of $(x - 4)^2 + (y + 3)^2 = 25$.

   answer:

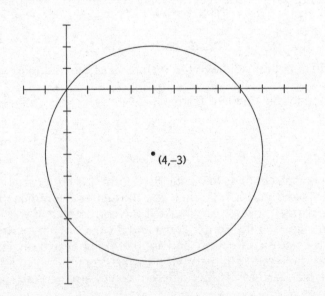

$(4,-3)$

The equation represents a circle. First plot the center, $(4, -3)$. This is a circle with radius 5, so count 5 units to the left, right, up, and down from that center. Mark these positions with dots. Then connect them.

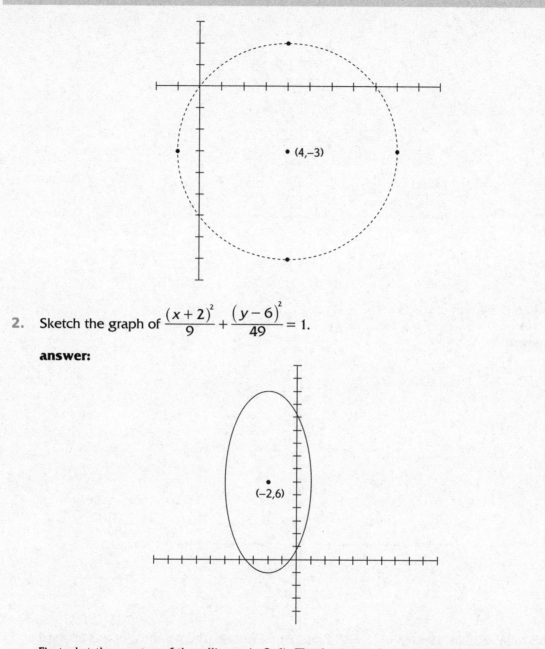

2. Sketch the graph of $\dfrac{(x+2)^2}{9} + \dfrac{(y-6)^2}{49} = 1$.

answer:

First plot the center of the ellipse, (−2,6). The horizontal axis is 6 units long—3 units in either direction from the center. The vertical axis is 14 units long—7 units in either direction from the center. Plot the points that would be the endpoints of the axes. Then connect them smoothly to form the ellipse.

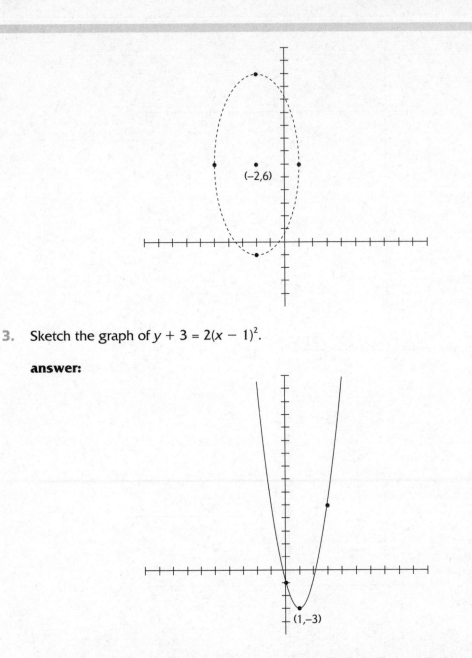

3. Sketch the graph of $y + 3 = 2(x - 1)^2$.

answer:

This parabola has a vertex of $(1, -3)$. Plot that point first. The parabola opens upward and is narrower because of the 2 multiplier. Find two other points on the parabola to help with the graph. Two possible points are $(0, -1)$ and $(3, 5)$.

4. Sketch the graph of $y^2 + 2y + 2x - 7 = 0$.

    answer:

    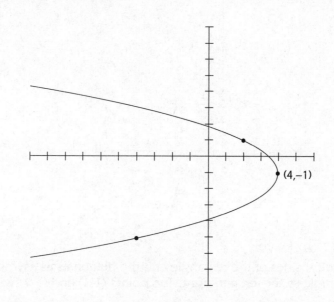

    (4,−1)

    This is the equation of a parabola, because there's only one squared term. Rewrite the equation in the standard form using completing the square to get $x - 4 = -\frac{1}{2}(y+1)^2$. The y term is squared, and the multiplier, $\frac{1}{2}$, is negative, so it opens to the left. The multiplier also makes the graph wider or flatter. Two other points on the graph, used to help with the sketch, might be $(-4,-5)$ and $(2,1)$.

5. Sketch the graph of $16x^2 - y^2 - 64x + 2y + 47 = 0$.

    answer:

    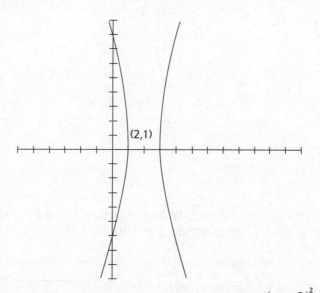

    (2,1)

    First, use completing the square to rewrite the equation as $\dfrac{(x-2)^2}{1} - \dfrac{(y-1)^2}{16} = 1$. You should know that this is an hyperbola, because the coefficients of the two squared terms are different signs. From the standard form, you see that the center is at $(2,1)$. The value of a is 1, and the value of b is 4. Draw a rectangle $2a$ or 2 units wide and $2b$ or 8 units high with $(2,1)$ in the center. Then draw lines along the diagonals of that rectangle.

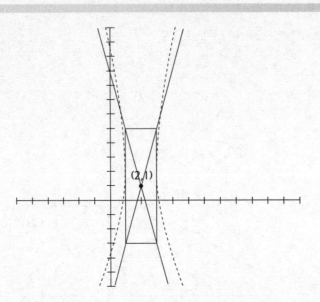

Now use the vertical sides of the rectangle and the diagonals as asymptotes to help sketch the hyperbola to the left and right. The points $(1,1)$ and $(3,1)$ will be points on the hyperbola.

6. Sketch the graph of $9x^2 + 25y^2 - 54x + 200y + 256 = 0$.

answer:

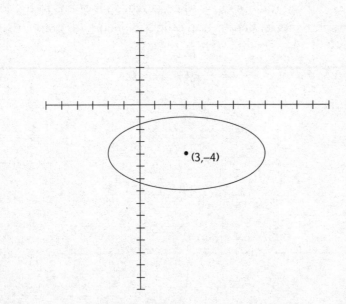

The coefficients of the squared terms are both positive, but they aren't the same value, so the graph is an ellipse. Change the equation to the standard form of the ellipse to get $\dfrac{(x-3)^2}{25} + \dfrac{(y+4)^2}{9} = 1$. The center is at $(3, -4)$. The horizontal axis is 10 units long, and the vertical axis is 6 units long. Starting from the center, count left and right 5 units and put points. Count up and down 3 units and put points. Draw in an ellipse from these points.

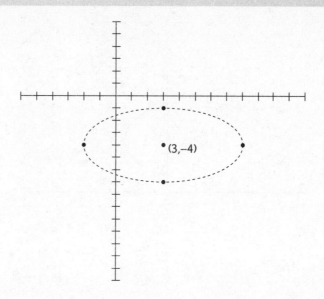

Work Problems

Use these problems to give yourself additional practice.

1. Sketch the graph of $(x - 7)^2 + (y - 3)^2 = 64$.

2. Sketch the graph of $\dfrac{(x + 2)^2}{25} + \dfrac{(y - 4)^2}{9} = 1$.

3. Sketch the graph of $y - 1 = -3(x + 2)^2$.

4. Sketch the graph of $2y^2 - 4y - x = 2$.

5. Sketch the graph of $100y^2 - 600y - 25x^2 - 250x = 1725$.

Worked Solutions

1.

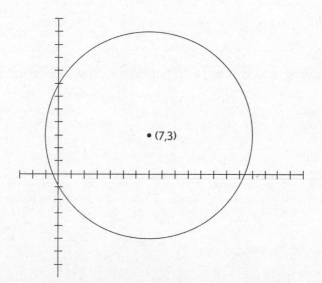

This is a circle with a center of (7,3) and a radius of 8 units.

2.

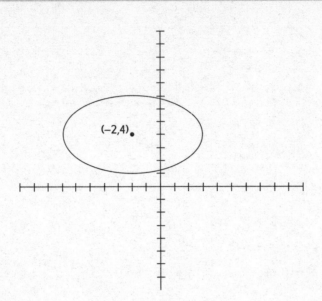

This is an ellipse with a center of (−2,4). The horizontal axis is 10 units and the vertical axis is 6 units.

3.

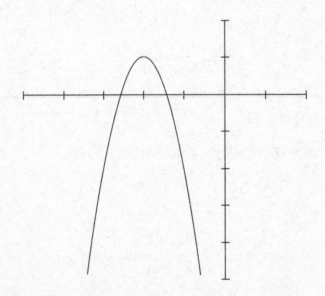

This is a parabola with its vertex at (−2,1). It opens downwards and is relatively narrow.

4.

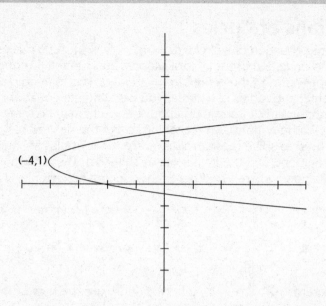

This parabola has a vertex of $(-4,1)$. The positive multiplier makes it open to the right.

5.

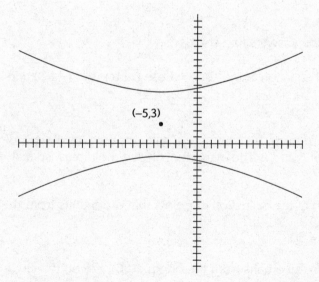

This hyperbola has a center at $(-5,3)$. The y^2 term is positive, so the hyperbola opens upward and downward.

Writing Equations of Circles

The standard form for the equation of a circle, $(x - h)^2 + (y - k)^2 = r^2$, makes it easy to determine the characteristics of the circle and use it in applications. You can tell, immediately, from this standard form what the coordinates of the center, (h,k), are and what the radius, r, is. If you're given the coordinates of a center and radius of a circle, you can write its equation. However, sometimes it's not quite that straightforward. Information about the circle may be more disguised, and you might have to do some computations and reasoning to get the desired result. Some other formulas that are frequently used to solve these problems are the formulas for the midpoint of a segment, the length of a segment, and the Pythagorean Theorem. These formulas, with explanations, are found in the Introduction and are listed here for your convenience.

Midpoint of a Segment $\quad M = \left(\dfrac{x_2 + x_1}{2}, \dfrac{y_2 + y_1}{2} \right)$ where (x_1,y_1) and (x_2,y_2) are the endpoints

Length of a Segment $\quad d = \sqrt{(x_2 - x_1)^2 + (y_2 - y_1)^2}$ where (x_1,y_1) and (x_2,y_2) are the endpoints

Pythagorean Theorem $\quad a^2 + b^2 = c^2$ where a and b are the legs of the right triangle and c is the hypotenuse

Example Problems

These problems show the answers and solutions.

1. Find the standard equation of the circle whose center is at $(-1,3)$ and whose diameter is 14.

   answer: $(x + 1)^2 + (y - 3)^2 = 49$

   Using the standard form, $(x - h)^2 + (y - k)^2 = r^2$, the coordinates of the center go right into the places for h and k. The diameter is twice the radius, so first divide that by 2 and then square the 7.

2. Find the equation of the collection of points that are 6 units from the origin.

   answer: $x^2 + y^2 = 36$

   The points that are all 6 units from the origin, $(0,0)$, are on the circle with a radius of 6. $(x - 0)^2 + (y - 0)^2 = 6^2$.

3. Find the equation of the circle that goes through the points $(-3,0)$ and $(5,-6)$ and whose center lies on the segment between the points.

   answer: $(x + 1)^2 + (y + 3)^2 = 25$

   A segment with endpoints on a circle that has the center lying on it is a diameter of the circle. The midpoint of the segment is the center of the circle. Using the formula for the midpoint of a segment, $M = \left(\dfrac{x_2 + x_1}{2}, \dfrac{y_2 + y_1}{2} \right)$, the center is $\left(\dfrac{-3 + 5}{2}, \dfrac{0 + (-6)}{2} \right) = (1, -3)$.
   The radius of the circle is the distance from the center to one of the points on the circle. Using the center, $(1,-3)$, and the point, $(-3,0)$ (either point will work and give the same answer), the distance between them is $d = \sqrt{(x_2 - x_1)^2 + (y_2 - y_1)^2} = \sqrt{(-3 - 1)^2 + (0 - (-3))^2}$
   $= \sqrt{(-4)^2 + 3^2} = \sqrt{16 + 9} = \sqrt{25} = 5$. A circle with a center of $(1,-3)$ and a radius of 5 has the standard equation $(x - 1)^2 + (y + 3)^2 = 5^2$.

4. Find the equation of the circle with a radius of 25, which also goes through the origin and has its center on the line $x = 24$.

    answer: $(x - 24)^2 + (y - 7)^2 = 625$ and $(x - 24)^2 + (y + 7)^2 = 625$

    There are two circles for which this is true. Draw the line $x = 24$ on coordinate axes, and then draw the radius of a circle that's 25 units long going from the origin to that line.

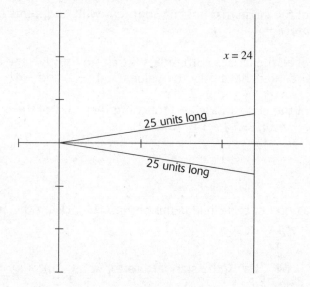

    There are two ways to do this. One radius is in the first quadrant, and the other is in the fourth quadrant. To find the centers of the circles, you need to find how high above and how low below the x-axis the points where the radius touches the vertical line are. Complete the right triangle that has a side measuring 24 units and a hypotenuse of 25 units by applying the Pythagorean Theorem: $a^2 + b^2 = c^2$ or $24^2 + b^2 = 25^2$; $576 + b^2 = 625$; $b^2 = 625 - 576 = 49$; $b = \sqrt{49} = 7$. The third side, the one going from the x-axis up to the point where the radius touches $x = 24$, is 7 units long. Also, in the fourth quadrant, the one going down measures 7 units. That makes the coordinates of the two points that can be centers $(24, 7)$ and $(24, -7)$. The circles have those coordinates for centers and a radius of 25 units.

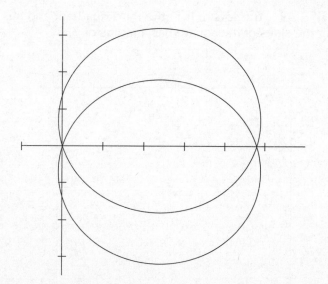

Work Problems

Use these problems to give yourself additional practice.

1. Find the equation of the circle whose points are all $\frac{1}{3}$ unit from the origin.

2. Find the equation of the circle whose points are all 8 units from the point $(-6,7)$.

3. Find the equation of the circle that has the segment with endpoints $(-5,0)$ and $(1,0)$ as endpoints of a diameter.

4. Find the equation of all the circles with radius 5 that go through the origin, and the coordinates of their centers have only the values 3, 4, -3, and -4.

5. Find the equation of the circle inside and touching the sides of the square that has vertices (corners) of $(0,0)$, $(-14,0)$, $(-14,-14)$, $(0,-14)$.

Worked Solutions

1. $x^2 + y^2 = \frac{1}{9}$ The center of this circle is the origin, $(0,0)$. Using the standard form, $(x-0)^2 + (y-0)^2 = \left(\frac{1}{3}\right)^2$.

2. $(x+6)^2 + (y-7)^2 = 64$ Using the standard form, $(x-(-6))^2 + (y-7)^2 = 8^2$.

3. $(x+2)^2 + y^2 = 9$ The center lies on the x-axis between $x = -5$ and $x = 1$. That segment is 6 units long, and the midpoint is at $\left(\frac{-5+1}{2}, \frac{0+0}{2}\right) = (-2,0)$. The 6-unit segment is the diameter, so the radius is 3 units. Using the standard form, $(x-(-2))^2 + (y-0)^2 = 3^2$.

4. $(x-3)^2 + (y-4)^2 = 25$, $(x-4)^2 + (y-3)^2 = 25$, $(x+3)^2 + (y-4)^2 = 25$, $(x-3)^2 + (y+4)^2 = 25$, $(x-4)^2 + (x+3)^2 = 25$, $(x+4)^2 + (y-3)^2 = 25$, $(x+3)^2 + (y+4)^2 = 25$, $(x+4)^2 + (y+3)^2 = 25$ The most familiar Pythagorean Triple is the 3,4,5 Pythagorean Triple. Using all possible combinations of 3's and 4's and -3's and -4's in all orders, you get eight different equations of circles.

5. $(x+7)^2 + (y+7)^2 = 49$ The square is in the third quadrant and has a center of $(-7,-7)$. A circle touching the sides of the square has a radius of 7.

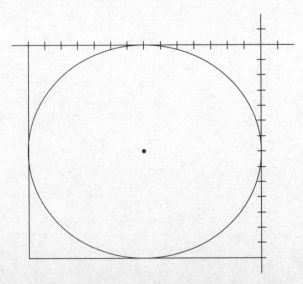

Chapter 9
Systems of Equations and Inequalities

Solving a system of equations means to find a common solution or common solutions to two or more equations. If the system consists of two lines, there will be one common solution or no common solution (if the lines are parallel) or an infinite number of common solutions (if the equations are of the same line). A line and a parabola can intersect in two or one or no places, so there can be two or one or no common solutions. As soon as you look for common solutions in curves and conics, you'll find possibilities for three or four or more common solutions. Each situation has to be considered independently.

Solving Linear Systems Using the Addition Method

When you're solving a system of *linear* equations, it means that you're trying to find the common solution of lines. One method you can use is a method called *linear combinations*, *elimination*, or more familiarly, *the addition method*. The goal is to add the two equations together and have the result give you one part of the answer—the value of one of the variables. This will happen if the other pair of variables in the two equations are opposites of one another, so they add up to 0. Sometimes it takes a little manipulating of one or more of the equations, such as multiplying through by some number, to make this work successfully.

Example Problems
These problems show the answers and solutions.

1. Solve the system $\begin{array}{l} 3x + 4y = 2 \\ 5x - 3y = 13 \end{array}$.

   answer: $x = 2$, $y = -1$ or $(2, -1)$

   Adding the two equations together will not result in anything productive. Either the x terms or the y terms need to be opposites of one another so that the addition will eliminate a term. Multiplying each term in the top equation by 3 and each term in the bottom equation by 4 will result in $+12y$ and $-12y$.

   $$9x + 12y = 6$$
   $$20x - 12y = 52$$

   Adding the two equations together results in the equation $29x = 58$. Solving for x by dividing each side by 29, $x = 2$. Now substitute the 2 for x into one of the *original* equations and solve for y: $3(2) + 4y = 2$, $6 + 4y = 2$, $4y = -4$, $y = -1$.

2. Solve the system $\begin{array}{l} y = 3x - 8 \\ 6x - 2y = 7 \end{array}$.

answer: No solution

To use the addition method, the two equations first have to be written in the same format. In this case, subtract $3x$ from each side in the first equation.

$$-3x + y = -8$$
$$6x - 2y = 7$$

Now multiply each term in the top equation by 2.

$$-6x + 2y = -16$$
$$6x - 2y = 7$$

Adding the two equations together, you get $0 + 0 = -9$. This is a false statement. There is no solution. This also means that the two lines are parallel to one another and will never cross.

Work Problems

Use these problems to give yourself additional practice.

1. Solve the system $\begin{array}{l} 4x - 9y = 23 \\ 2x - 4y = 10 \end{array}$.

2. Solve the system $\begin{array}{l} 5x + 3y = 7 \\ 3x + 2y = 5 \end{array}$.

3. Solve the system $\begin{array}{l} 4x + 5y = 32 \\ 2y - 6x = 9 \end{array}$.

4. Solve the system $\begin{array}{l} 8x - 4y = 7 \\ 4x - 2y = 3 \end{array}$.

5. Solve the system $\begin{array}{l} x + 6y = -14 \\ 3x - y = 34 \end{array}$.

Worked Solutions

1. **$x = -1$, $y = -3$ or $(-1,-3)$**   Multiply each term in the bottom equation by -2 to get
    $$4x - 9y = 23$$
    $$-4x + 8y = -20$$

    Adding the two equations together, you get $-y = 3$, so $y = -3$. Substitute that value in for the y in the first equation to get $4x - 9(-3) = 23$, $4x + 27 = 23$, $4x = -4$ or $x = -1$.

2. **$x = -1$, $y = 4$ or $(-1,4)$**   Multiply each term in the top equation by 2 and each term in the bottom equation by -3. That gives you $\begin{array}{l} 10x + 6y = 14 \\ -9x - 6y = -15 \end{array}$. Adding the two equations together, you get $x = -1$. Substitute that back into the *original* first equation to solve for y: $5(-1) + 3y = 7$, $-5 + 3y = 7$, $3y = 12$, $y = 4$.

3. $x = \frac{1}{2}$, $y = 6$ or $\left(\frac{1}{2}, 6\right)$ First rearrange the terms on the left so that the x terms and y terms line up under one another. That gives you the system $\begin{array}{l} 4x + 5y = 32 \\ -6x + 2y = 9 \end{array}$. Now multiply each term in the top equation by 3 and each term in the bottom equation by 2: $\begin{array}{l} 12x + 15y = 96 \\ -12x + 4y = 18 \end{array}$. Add the two equations together and solve for y: $19y = 114$, $y = 6$. Substitute 6 for y in the *original* first equation and solve for x:

$4x + 5(6) = 32$, $4x + 30 = 32$, $4x = 2$, $x = \frac{1}{2}$.

4. **No solution**   Multiply the bottom equation through by -2: $\begin{array}{l} 8x - 4y = 7 \\ -8x + 4y = -6 \end{array}$. Adding the two equations together, you get a contradiction, $0 + 0 = 1$. There is no possible solution.

5. $x = 10$, $y = -4$ or $(10, -4)$ Multiply the bottom equation through by 6: $\begin{array}{l} x + 6y = -14 \\ 18x - 6y = 204 \end{array}$. Add the two equations together and solve for x: $19x = 190$, $x = 10$. Substitute 10 for the x in the first equation. $10 + 6y = -14$, $6y = -24$, $y = -4$.

Solving Linear Equations Using Substitution

A method used to solve both linear equations and nonlinear equations is the substitution method. The addition method will always work for linear equations, but it frequently does not work with nonlinear equations. The substitution method is usually not the method of choice with linear equations unless it's convenient to solve for one of the variables and not create a fraction. The substitution method is just that: You substitute the equivalence of one of the variables into the other equation and then solve for a variable. Then you finish by substituting the value found back in to get the other variable.

Example Problems

These problems have the answers and solutions.

1. Solve the system $\begin{array}{l} 3x - 7y = 29 \\ x - 3y = 11 \end{array}$ using substitution.

   answer: $x = 5$, $y = -2$ or $(5, -2)$

   Solve the second equation for x by adding $3y$ to each side of the equation. It's the only variable with a coefficient of 1, so you won't have to divide through by some number and create fractions: $x - 3y = 11$ becomes $x = 3y + 11$. Now substitute the $3y + 11$ for the x in the first equation. When you solve for a variable in one equation, be sure to substitute back into the *other* equation: $3(3y + 11) - 7y = 29$; $9y + 33 - 7y = 29$; $2y + 33 = 29$; $2y = -4$; $y = -2$. The quickest way to solve for x is to go to the equation $x = 3y + 11$. Letting the y be equal to -2, $x = 3(-2) + 11 = -6 + 11 = 5$.

2. Solve the system $\begin{array}{l} 9x + 4y = 35 \\ 4x - y = 10 \end{array}$ using substitution.

   answer: $x = 3$, $y = 2$ or $(3, 2)$

   Solve for y in the second equation by subtracting 10 from each side and adding y to each side: $4x - y = 10$, $4x - 10 = y$. Substitute $4x - 10$ into the first equation, replacing the y: $9x + 4(4x - 10) = 35$; $9x + 16x - 40 = 35$; $25x = 75$; $x = 3$. Because $y = 4x - 10$, when you replace the x with 3, you get the value of y: $y = 4(3) - 10 = 2$.

Work Problems

Use these problems to give yourself additional practice. Solve each system using substitution.

1. $5x - 4y - 11 = 0$
    $2x - y = 2$

2. $3x + 2y = 15$
    $x + y = 6$

3. $6x - 2y = 6$
    $3x = y + 3$

4. $x - 4y + 5 = 0$
    $2x - 3y = 0$

5. $15x + 14y = 12$
    $5x + y + 7 = 0$

Worked Solutions

1. **$x = -1$, $y = -4$ or $(-1, -4)$**   Solve the second equation for y by adding y to each side and subtracting 2 from each side: $y = 2x - 2$. Substitute $2x - 2$ in for the y in the first equation and solve for x: $5x - 4(2x - 2) - 11 = 0$, $5x - 8x + 8 - 11 = 0$, $-3x - 3 = 0$, $-3x = 3$, $x = -1$. Now replace the x with -1 in $y = 2x - 2$ to get $y = 2(-1) - 2 = -2 - 2 = -4$.

2. **$x = 3$, $y = 3$ or $(3, 3)$**   Solve the second equation for x (you could also solve it for y) to get $x = 6 - y$. Substitute $6 - y$ for x in the first equation and solve for y: $3(6 - y) + 2y = 15$, $18 - 3y + 2y = 15$, $-y = -3$, $y = 3$. Now substitute the 3 into $x = 6 - y$ to get $x = 6 - 3 = 3$.

3. **All values satisfying either equation**   Solve for y in the second equation by subtracting 3 from each side: $y = 3x - 3$. Substitute $3x - 3$ into the first equation to get $6x - 2(3x - 3) = 6$. From this $6x - 6x + 6 = 6$, $0 = 0$. This isn't a contradiction, like you find in the problems where there is no solution. Rather, this is *always* true. This occurs when the two equations represent the same line. So any values that satisfy the first equation will also satisfy the second. There are an infinite number of correct solutions. All will be of the form x and $y = 3x - 3$ or $(x, 3x - 3)$.

4. **$x = 3$, $y = 2$ or $(3, 2)$**   Solve the first equation for x: $x = 4y - 5$. Substitute the $4y - 5$ for x in the second equation to get $2(4y - 5) - 3y = 0$. Now solve for y. $8y - 10 - 3y = 0$, $5y = 10$, $y = 2$. Replace the y with 2 in $x = 4y - 5$ to get $x = 4(2) - 5 = 8 - 5 = 3$.

5. **$x = -2$, $y = 3$ or $(-2, 3)$**   Solve the second equation for y to get $y = -5x - 7$. Replace the y in the first equation with $-5x - 7$ to get $15x + 14(-5x - 7) = 12$. Solve for x: $15x - 70x - 98 = 12$, $-55x = 110$, $x = -2$. Because $y = -5x - 7$, then $y = -5(-2) - 7 = 10 - 7 = 3$.

Solving Linear Equations Using Cramer's Rule

Cramer's Rule for solving linear equations isn't necessarily any easier or more efficient than using the addition method or substitution, but it's a method that can be programmed into a computer or graphing calculator. In fact, it's based on *determinants*, which are covered in Chapter 10. Use a method like this if there are a lot of systems to solve and if there are apt to be a lot of fractions and decimals in the answers. Computers and graphing calculators can help you through some of the more tedious computations.

Begin with a system of two linear equations, written in the form $Ax + By = C$ and $Dx + Ey = F$. The coefficients and constants are the A, B, C, D, E, and F, and they are what are used in this rule.

The solution for the system $\begin{matrix} Ax + By = C \\ Dx + Ey = F \end{matrix}$ is $x = \dfrac{CE - BF}{AE - BD}$, $y = \dfrac{AF - CD}{AE - BD}$. This may look a bit complicated, but one way to help you remember the products and differences is to look at the square of coefficients $\begin{matrix} A & B \\ D & E \end{matrix}$. Multiply diagonally $A{\cdot}E$ and $B{\cdot}D$. The difference of those two products goes in the denominator of each part of the solution. The numerators also have the differences of diagonal products, but in each case two of the coefficients are replaced by the constants. When solving for x, replace the A and D (the coefficients of the x's) with C and F to get $\begin{matrix} C & B \\ F & E \end{matrix}$. Now the diagonal products are $C{\cdot}E$ and $B{\cdot}F$. The difference between these two products is found in the numerator of the x part of the solution. When solving for y, replace the B and E (the coefficients of the y's) with C and F to get $\begin{matrix} A & C \\ D & F \end{matrix}$. Now the diagonal products are $A{\cdot}F$ and $C{\cdot}D$. The difference between these two products is found in the numerator of the y part of the solution.

Example Problems

These problems show the answers and solutions.

1. Solve the system $\begin{matrix} 2x + 3y = 17 \\ 6x - 5y = 9 \end{matrix}$ using Cramer's Rule.

    answer: $x = 4$, $y = 3$ or $(4,3)$

    The denominator of each part of the solution is formed from the square $\begin{matrix} 2 & 3 \\ 6 & -5 \end{matrix}$ giving you $2(-5) - 3(6) = -10 - 18 = -28$. The numerator of the x part of the solution is formed from the square of numbers $\begin{matrix} 17 & 3 \\ 9 & -5 \end{matrix}$. The numerator is $17(-5) - 3(9) = -85 - 27 = -112$. The value of x is $\dfrac{-112}{-28} = 4$. The numerator of the y part of the solution is formed from the square of numbers $\begin{matrix} 2 & 17 \\ 6 & 9 \end{matrix}$. The numerator is $2(9) - 17(6) = 18 - 102 = -84$. The value of y is $\dfrac{-84}{-28} = 3$.

2. Solve the system $\begin{matrix} 3x + 2y = 7 \\ 5x - 3y = 1 \end{matrix}$ using Cramer's Rule.

    answer: $x = \dfrac{23}{19}$, $y = \dfrac{32}{19}$ or $\left(\dfrac{23}{19}, \dfrac{32}{19} \right)$

    The denominator of each part of the solution is formed from the square $\begin{matrix} 3 & 2 \\ 5 & -3 \end{matrix}$ giving you $3(-3) - 2(5) = -9 - 10 = -19$. The numerator of the x part of the solution is formed from the square of numbers $\begin{matrix} 7 & 2 \\ 1 & -3 \end{matrix}$, so the numerator is $7(-3) - 2(1) = -21 - 2 = -23$. The numerator of the y part of the solution is formed from the square of numbers $\begin{matrix} 3 & 7 \\ 5 & 1 \end{matrix}$, so the numerator is $3(1) - 7(5) = 3 - 35 = -32$. The solution for the system is $x = \dfrac{-23}{-19} = \dfrac{23}{19}$, $y = \dfrac{-32}{-19} = \dfrac{32}{19}$. Problems like this are more easily handled using Cramer's Rule than using the addition method or substitution. The fractions can get very messy.

Work Problems

Use these problems to give yourself additional practice. Solve each system using Cramer's Rule.

1. $x - 4y = 14$
   $x + y = 10$

2. $4x - 5y = 11$
   $3x - 7y = 2$

3. $5x + y = 7$
   $4x + y = 16$

4. $13x + 2y = 8$
   $9x - 6y = 1$

5. $100x - y = 10$
   $300x + 4y = 20$

Worked Solutions

1. $x = \dfrac{54}{5}$, $y = -\dfrac{4}{5}$ or $\left(\dfrac{54}{5}, -\dfrac{4}{5}\right)$ The following are the number squares used. For the

   denominator: $\begin{matrix} 1 & -4 \\ 1 & 1 \end{matrix}$; for the numerator of x: $\begin{matrix} 14 & -4 \\ 10 & 1 \end{matrix}$; for the numerator of y: $\begin{matrix} 1 & 14 \\ 1 & 10 \end{matrix}$.

   So $x = \dfrac{14(1) - (-4)(10)}{1(1) - (-4)(1)} = \dfrac{14 + 40}{1 + 4} = \dfrac{54}{5}$, and $y = \dfrac{1(10) - (14)(1)}{1(1) - (-4)(1)} = \dfrac{10 - 14}{1 + 4} = \dfrac{-4}{5}$.

2. $x = \dfrac{67}{13}$, $y = \dfrac{25}{13}$ or $\left(\dfrac{67}{13}, \dfrac{25}{13}\right)$ The following are the number squares used. For the

   denominator: $\begin{matrix} 4 & -5 \\ 3 & -7 \end{matrix}$; for the numerator of x: $\begin{matrix} 11 & -5 \\ 2 & -7 \end{matrix}$; for the numerator of y: $\begin{matrix} 4 & 11 \\ 3 & 2 \end{matrix}$.

   So $x = \dfrac{11(-7) - (-5)(2)}{4(-7) - (-5)(3)} = \dfrac{-77 + 10}{-28 + 15} = \dfrac{-67}{-13} = \dfrac{67}{13}$, and

   $y = \dfrac{4(2) - (11)(3)}{4(-7) - (-5)(3)} = \dfrac{8 - 33}{-28 + 15} = \dfrac{-25}{-13} = \dfrac{25}{13}$.

3. $x = -9$, $y = 52$ or $(-9, 52)$ The following are the number squares used. For the

   denominator: $\begin{matrix} 5 & 1 \\ 4 & 1 \end{matrix}$; for the numerator of x: $\begin{matrix} 7 & 1 \\ 16 & 1 \end{matrix}$; for the numerator of y: $\begin{matrix} 5 & 7 \\ 4 & 16 \end{matrix}$. So

   $x = \dfrac{7(1) - (1)(16)}{5(1) - (1)(4)} = \dfrac{7 - 16}{5 - 4} = \dfrac{-9}{1} = -9$, and $y = \dfrac{5(16) - (7)(4)}{5(1) - (1)(4)} = \dfrac{80 - 28}{5 - 4} = \dfrac{52}{1} = 52$.

4. $x = \dfrac{25}{48}$, $y = \dfrac{59}{96}$ or $\left(\dfrac{25}{48}, \dfrac{59}{96}\right)$ The following are the number squares used.

   For the denominator: $\begin{matrix} 13 & 2 \\ 9 & -6 \end{matrix}$; for the numerator of x: $\begin{matrix} 8 & 2 \\ 1 & -6 \end{matrix}$; for the numerator

   of y: $\begin{matrix} 13 & 8 \\ 9 & 1 \end{matrix}$. So $x = \dfrac{8(-6) - (2)(1)}{13(-6) - (2)(9)} = \dfrac{-48 - 2}{-78 - 18} = \dfrac{-50}{-96} = \dfrac{25}{48}$,

   and $y = \dfrac{13(1) - 9(8)}{13(-6) - 2(9)} = \dfrac{13 - 72}{-78 - 18} = \dfrac{-59}{-96} = \dfrac{59}{96}$.

5.　　$x = \dfrac{3}{35}$, $y = -\dfrac{10}{7}$ or $\left(\dfrac{3}{35}, -\dfrac{10}{7}\right)$　The following are the number squares used. For the

denominator: $\begin{matrix} 100 & -1 \\ 300 & 4 \end{matrix}$; for the numerator of x : $\begin{matrix} 10 & -1 \\ 20 & 4 \end{matrix}$; for the numerator of y: $\begin{matrix} 100 & 10 \\ 300 & 20 \end{matrix}$.

$$\text{So } x = \frac{10(4) - (-1)(20)}{100(4) - (-1)(300)} = \frac{40 + 20}{400 + 300} = \frac{60}{700} = \frac{3}{35}, \text{ and}$$

$$y = \frac{100(20) - (10)(300)}{100(4) - (-1)(300)} = \frac{2000 - 3000}{400 + 300} = \frac{-1000}{700} = -\frac{10}{7}$$

Systems of Non-Linear Equations

When equations representing lines and equations representing conics are solved for their common solutions, one of three things can happen. There can be two solutions, one solution, or no solution. When two conics are solved for a common solution, there can be as many as four different solutions. Look at this situation that occurs when a parabola and circle intersect.

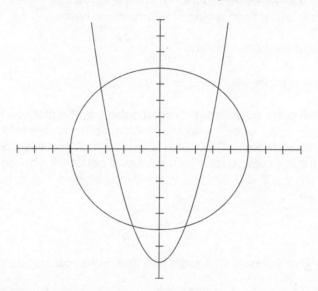

There are four separate points where the two curves cross. These represent the four different solutions you find when you solve the system of equations.

The substitution method is usually used to solve these systems, because it will always work. Occasionally, you can use the addition method, if it results in eliminating a variable and all its powers completely.

Example Problems

These problems show the answers and solutions.

1.　Solve for the solution(s) of the system $\begin{aligned} y &= x^2 - 5 \\ x^2 + y^2 &= 25 \end{aligned}$.

　　answer: $x = 0$, $y = -5$ and $x = 3$, $y = 4$ and $x = -3$, $y = 4$ or $(0,-5)$, $(3,4)$, $(-3,4)$

　　Use substitution to do this problem. Solve for x^2 in the first equation to get $x^2 = y + 5$. Replace the x^2 in the second equation with $y + 5$ and then solve the quadratic equation that results; $x^2 + y^2 = 25$ becomes $y + 5 + y^2 = 25$, $y^2 + y - 20 = 0$. This quadratic factors: $(y + 5)(y - 4) = 0$. The two solutions are $y = -5$ and $y = 4$. Substitute those values back into the first equation—the parabola. You are less likely to end up with extraneous solutions if you go back to the equation with lower powers; in this case just the y has a lower power. When $y = -5$, $-5 = x^2 - 5$, $x^2 = 0$, $x = 0$. When $y = 4$, $4 = x^2 - 5$, $x^2 = 9$, $x = \pm 3$. Pair up the x's and the y's accordingly.

2. Solve for the solution(s) of the system $\begin{aligned} x^2 + y^2 &= 58 \\ 3x^2 + y^2 &= 156 \end{aligned}$.

    answer: $x = 7$, $y = 3$ and $x = 7$, $y = -3$ and $x = -7$, $y = 3$ and $x = -7$, $y = -3$ or $(7,3)$, $(7,-3)$, $(-7,3)$, $(-7,-3)$

    This is a case in which the addition method works. Multiply each term in the first equation by -1 and then add the two equations together: $\begin{aligned} -x^2 - y^2 &= -58 \\ 3x^2 + y^2 &= 156 \end{aligned}$. Adding, you get $2x^2 = 98$, $x^2 = 49$, $x = \pm 7$. Substituting first 7 and then -7 into the *original* first equation, $7^2 + y^2 = 58$, $49 + y^2 = 58$, $y^2 = 9$, $y = \pm 3$, $(-7)^2 + y^2 = 58$, $49 + y^2 = 58$, $y = \pm 3$. Pairing the two values for x with each of their two corresponding y values, you get the four different solutions. Graphing this circle and ellipse would illustrate the solutions.

3. Solve for the solution(s) of the system $\begin{aligned} y &= 5 + 4x - x^2 \\ x + y &= 9 \end{aligned}$.

    answer: $x = 1$, $y = 8$ and $x = 4$, $y = 5$ or $(1,8)$, $(4,5)$

    Solve for y in the second equation and replace the y in the first equation with its equivalent. $y = 9 - x$, $9 - x = 5 + 4x - x^2$. Move all the terms to the left and solve the quadratic equation: $x^2 - 5x + 4 = 0$, $(x - 1)(x - 4) = 0$. The two solutions here are $x = 1$ and $x = 4$. Substituting these back into the second equation, you get the corresponding y values: $1 + y = 9$, $y = 8$ and $4 + y = 9$, $y = 5$.

Work Problems

Use these problems to give yourself additional practice. Solve each system for the solution(s).

1. $\begin{aligned} x^2 + y^2 &= 40 \\ x^2 - 5y^2 &= 16 \end{aligned}$

2. $\begin{aligned} y &= x^2 - 3x + 2 \\ x - y + 2 &= 0 \end{aligned}$

3. $\begin{aligned} x^2 + y^2 &= 100 \\ 16x + 160 &= y^2 \end{aligned}$

4. $\begin{aligned} y &= 4 - x^2 \\ 2y &= x + 10 \end{aligned}$

5. $\begin{aligned} x^2 + y^2 &= 10 \\ 3x^2 + 2y^2 &= 21 \end{aligned}$

Worked Solutions

1. $x = 6$, $y = 2$ and $x = 6$, $y = -2$ and $x = -6$, $y = 2$ and $x = -6$, $y = -2$ or **(6,2), (6, −2), (−6,2), (−6,−2)**   Multiply each term in the bottom equation by -1 and add the two equations together: $\begin{aligned} x^2+y^2 &= 40 \\ -x^2+5y^2 &=-16 \end{aligned}$. Adding gives you $6y^2 = 24$, $y^2 = 4$, $y = \pm2$. Substituting the two values for y into the first equation, $y = 2$, $x^2 + 2^2 = 40$, $x^2 = 36$, $x = \pm6$ and $y = -2$, $x^2 + (-2)^2 = 40$, $x^2 = 36$, $x = \pm6$.

2. $x = 0$, $y = 2$ and $x = 4$, $y = 6$ or **(0,2),(4,6)**   Substitute the equivalent of y from the first equation into the second equation to get $x - (x^2 - 3x + 2) + 2 = 0$; $x - x^2 + 3x - 2 + 2 = 0$. Simplify and move all the terms to the right: $x^2 - 4x = 0$. This factors into $x(x - 4) = 0$ and gives the solutions $x = 0$ and $x = 4$. Substitute back into the first equation, $y = 0^2 - 3(0) + 2 = 2$ and $y = 4^2 - 3(4) + 2 = 16 - 12 + 2 = 6$.

3. $x = -10$, $y = 0$ and $x = -6$, $y = 8$ and $x = -6$, $y = -8$ or **(−10,0), (−6,8), (−6,−8)**   Substitute the equivalence of y^2 from the second equation into the first equation to get $x^2 + 16x + 160 = 100$. Subtract 100 from each side and then factor the resulting quadratic equation: $x^2 + 16x + 60 = (x + 10)(x + 6) = 0$. This gives you the two solutions $x = -10$ and $x = -6$. Substituting the -10 back into the second equation, $16(-10) + 160 = y^2$, $-160 + 160 = y^2$, $0 = y^2$, $y = 0$. Now, substituting the -6 back into the second equation, $16(-6) + 160 = y^2$, $-96 + 160 = y^2$, $64 = y^2$, $y = \pm8$.

4. **No solution**   Substitute the equivalent of y from the first equation into the second equation to get $2(4 - x^2) = x + 10$, $8 - 2x^2 = x + 10$. Move all the terms to the right and solve the quadratic equation. $2x^2 + x + 2 = 0$ doesn't factor, so, using the quadratic formula, $x = \dfrac{-1 \pm \sqrt{1^2 - 4(2)(2)}}{2(2)} = \dfrac{-1 \pm \sqrt{1 - 16}}{4}$. This results in a negative under the radical, so there's no real solution. The parabola and the line don't intersect.

5. $x = 1$, $y = 3$ and $x = 1$, $y = -3$ and $x = -1$, $y = 3$ and $x = -1$, $y = -3$ or **(1,3), (1, −3), (−1, 3), (−1,−3)**   Multiply each term in the first equation by -2, add the equations together, and solve for y. $\begin{aligned} -2x^2 - 2y^2 &=-20 \\ 3x^2 + 2y^2 &= 21 \end{aligned}$ gives you $x^2 = 1$, $x = \pm1$. Substituting 1 for x into the first equation, $1^2 + y^2 = 10$, $y^2 = 9$, $y = \pm3$. Substituting -1 for x gives the same two solutions.

Story Problems Using Systems of Equations

Practical applications or story problems are an important part of any study of algebra, because they illustrate some of the practical uses of algebra. Using systems of equations enables you to solve problems that involve more than one quantity or variable. The best way to illustrate what this means is to show some examples.

Example Problems

These problems show the answers and solutions.

1. Clark has 16 of a particular large coin and 12 of another, smaller, coin; he has a total of $5.20. Lois has a collection of the same types of coins as Clark. She has 5 of the larger coins and 22 of the smaller and has a total of $3.45. What denominations are these coins?

answer: Larger coin is a quarter; smaller coin is a dime.

Let the value of each of the larger coins, in cents, be represented by the variable x and the value of each of the smaller coins, in cents, be represented by the variable y. Then Clark's collection of coins can be described with $16x + 12y = 5.20$, and Lois' collection can be described with $5x + 22y = 3.45$. Either the addition method or Cramer's rule would work here. Using addition, multiply each term in the first equation by 5 and each term in the second equation by -16:

$\begin{aligned} 80x + 60y &= 26.00 \\ -80x - 352y &= -55.20 \end{aligned}$. Adding the equations together, $-292y = -29.20$. Divide each side by -292 to solve for y: $y = \dfrac{-29.20}{-292} = 0.10$. This is a dime. Substitute 0.10 for y in the first *original* equation to get $16x + 12(0.10) = 5.20$, $16x + 1.2 = 5.2$, $16x = 4$, $x = 0.25$. This is a quarter.

2. The height of a ball thrown up in the air can be determined by $s = -16t^2 + 128t + 144$ where s is the height of the ball in feet, and t is the number of seconds since it's thrown in the air. When will the ball be 384 feet in the air, and when will it hit the ground?

    answer: It'll be 384 feet high after 3 seconds (on the way up) and then, again, after 5 seconds (on the way down); it'll hit the ground after 9 seconds.

    To determine when it'll be 384 feet high, replace the s, the height, with 384 to get the quadratic equation $384 = -16t^2 + 128t + 144$. Subtract 384 from each side and then factor the quadratic: $0 = -16t^2 + 128t - 240$; $0 = -16(t^2 - 8t + 15) = -16(t - 3)(t - 5)$. The solutions to this quadratic are $t = 3$ and $t = 5$. To determine when the ball will hit the ground, solve for when $s = 0$. Substituting, $0 = -16t^2 + 128t + 144$. This factors into $0 = -16(t^2 - 8t - 9) = -16(t - 9)(t + 1)$. The solutions to this quadratic are $t = 9$ and $t = -1$. The 9 is the answer you want. The solution $t = -1$ doesn't really make sense, because you can't turn time backward. The reason this shows up as an answer is because the ball was thrown from a height of 144 feet in the air—perhaps from the top of a building—and fell to the ground.

Work Problems

Use these problems to give yourself additional practice.

1. Gloria is putting together a mixture of malted milk balls and chocolate covered raisins to make a bridge mix. The malted milk balls cost $2.25 per pound, and the chocolate covered raisins cost $3.75 per pound. What is the maximum number of pounds of each she should use to make a mixture consisting of a total of 20 pounds and costing no more than $51.00?

2. Randy has $450 in $5 bills and $10 bills. He has exactly 65 pieces of paper currency. How many of these are $10 bills?

3. A right triangle has a hypotenuse of length 75 centimeters. The sum of the measures of its sides is 168 centimeters. How long are the two legs?

4. A star player for the Peoria Chiefs Baseball team is doing well this season. So far, the sum of his home runs and total number of hits (including the home runs) is 99. The square of the number of home runs less the number of hits is 243. How many home runs does he have?

5. The sum of the squares of two positive numbers is 1885, and the difference of the squares of the numbers is 1643. What is the larger of the two numbers?

Worked Solutions

1. **16 pounds of malted milk balls and 4 pounds of chocolate covered raisins** Let m represent the number of pounds of malted milk balls and c represent the number of pounds of chocolate covered raisins. The total number of pounds is to be 20, so $m + c = 20$. Multiplying the number of pounds of each by the respective prices, $2.25m + 3.75c$ is the amount of money that will be spent. In order to maximize the number of pounds used, aim for using all \$51, so let $2.25m + 3.75c = 51.00$.
    The system to solve is $\begin{matrix} m + c = 20 \\ 2.25m + 3.75c = 51.00 \end{matrix}$. Multiply the terms in the first equation by -2.25. Adding the two equations $\begin{matrix} -2.25m - 2.25c = -45 \\ 2.25m + 3.75c = 51.00 \end{matrix}$, you get $1.50c = 6$, $c = 4$. If 4 is the number of pounds of chocolate covered raisins and you need a total of 20 pounds, that leaves 16 pounds for the malted milk balls.

2. **25 are 10 dollar bills** Let f represent the number of \$5 bills and t represent the number of \$10 bills. The total number of bills is 65, so $f + t = 65$. The total amount of money is \$450, so add the products of the numbers of bills and their worth to get $5f + 10t = 450$.
    The system of equations is $\begin{matrix} f + t = 65 \\ 5f + 10t = 450 \end{matrix}$. Multiply the terms in the first equation by -5 to get $\begin{matrix} -5f + -5t = -325 \\ 5f + 10t = 450 \end{matrix}$. Adding the equations together and solving for t, $5t = 125$, $t = 25$.

3. **The legs measure 21 centimeters and 72 centimeters** Let the lengths of the three sides of the right triangle be a, b, and c, where c is the length of the hypotenuse. The sum of the lengths of the sides is 168, so $a + b + 75 = 168$, $a + b = 93$. The Pythagorean Theorem applies to the lengths of the sides of a right triangle: $a^2 + b^2 = c^2$. So $a^2 + b^2 = 75^2 = 5625$.
    The system of equations is $\begin{matrix} a + b = 93 \\ a^2 + b^2 = 5625 \end{matrix}$. Solve for a in the first equation and substitute into the second equation: $a = 93 - b$. So $(93 - b)^2 + b^2 = 5625$. Then $8649 - 186b + b^2 + b^2 = 5625$. Simplifying and moving all the terms to the left, $2b^2 - 186b + 3024 = 0$. Divide each term by 2 to get $b^2 - 93b + 1512 = 0$. This factors into $(b - 21)(b - 72) = 0$, and the solutions are 21 and 72, the lengths of the other two sides.

4. **18 home runs** Let r represent the number of home runs and h represent the number of hits. The sum of the home runs and total number of hits is then $r + h = 99$. The square of the number of home runs less the number of hits can be represented by $r^2 - h$ and that is equal to 243. The system of equations to solve is $\begin{matrix} r + h = 99 \\ r^2 - h = 243 \end{matrix}$. Add the two equations together to get $r^2 + r = 342$, which is a quadratic equation. The quadratic $r^2 + r - 342 = 0$ can be factored into $(r - 18)(r + 19) = 0$. Only the solution $r = 18$ makes sense (you can't have a negative number of home runs). That means he had 18 home runs, or a total of 81 hits.

5. **The larger number is 42** Let x and y represent the two positive numbers. The sum of their squares is $x^2 + y^2$, which is equal to 1885. The difference of their squares is $x^2 - y^2$, which is equal to 1643. The system of equations to be solved is $\begin{matrix} x^2 + y^2 = 1885 \\ x^2 - y^2 = 1643 \end{matrix}$. Add the two together to get $2x^2 = 3528$, $x^2 = 1764$. Taking the square root of each side, $x = \pm 42$. You only want the positive number, so substitute 42 for x in the first equation to solve for y: $42^2 + y^2 = 1885$, $y^2 = 1885 - 1764 = 121$, $y = 11$. The larger of the two numbers is the 42.

Systems of Inequalities

An inequality statement can have an infinite number of solutions. If you write $x + y > 4$, then all of the points to the right and above the line $x + y = 4$ are solutions of the inequality. Put two inequalities together, and you can determine the solutions that they have in common—the solutions that work for both inequalities at the same time.

The process used to solve these systems of inequalities is to graph the inequalities and test for the solutions of each. Then determine where the solutions are shared or the same. When graphing a single inequality, you usually shade in the side of the line or curve that contains all the solutions. When graphing a system of inequalities, shade in the separate solutions and determine where the shading overlaps. That's your solution.

Example Problems

The following problems show the answers and the solutions.

1. Solve the system of inequalities $\begin{array}{l} 4x + y > 6 \\ x - y \le 9 \end{array}$.

   answer:

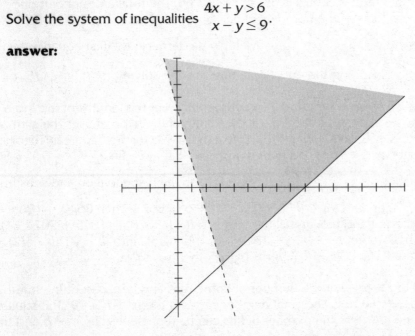

The inequality $4x + y > 6$ is graphed with a dashed line that goes through the y-axis at $(0,6)$ and goes downward with a slope of -4. See Chapter 1 for information on slopes and the intercepts of lines. The right side of the line is shaded, because points such as $(4,4)$ and $(6,7)$ are parts of the solution, and they lie to the right and above the line. The inequality $x - y \le 9$ is graphed with a solid line that goes through the y-axis at $(0,-9)$ and has a slope of 1. The left side, above the line, is shaded. The two lines intersect at the point $(3,-6)$. The intersection of the solutions of the inequalities is the region above and between the lines, as shown. Any point in that region is a solution of either inequality and a solution of the system—both inequalities.

2.   Solve the system of inequalities $\begin{array}{l} x - y \geq 3 \\ x^2 + y^2 \leq 25 \end{array}$.

answer:

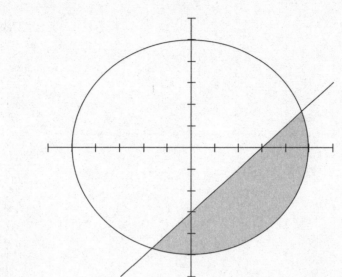

The inequality $x - y \geq 3$ is graphed as a solid line with a y-intercept of -3 and a slope of 1, and then the region under and to the right is shaded. The inequality $x^2 + y^2 \leq 25$ is graphed as the circle $x^2 + y^2 = 25$ with the interior shaded in. The two inequalities overlap inside the circle and under (to the right of) the line. A sample solution is the point $(4, -1)$.

Work Problems

Use these problems to give yourself additional practice. In each case, solve for the solution to the system of inequalities.

1. $\begin{array}{l} x - 3y \geq 4 \\ 2x + y \geq 8 \end{array}$

2. $\begin{array}{l} y \geq x^2 - 3 \\ y \leq 4 - x^2 \end{array}$

3. $\begin{array}{l} x \geq y^2 + 2y - 1 \\ y \leq 3 - x \end{array}$

4. $\begin{array}{l} 3x^2 + 4y^2 \geq 7 \\ y \leq x \end{array}$

5. $\begin{array}{l} 2x^2 + 5y^2 \leq 11 \\ x^2 + y^2 \geq 4 \end{array}$

Worked Solutions

1.

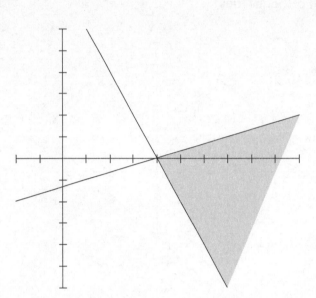

The inequality $x - 3y \geq 4$ is graphed with the line intersecting the y-axis at $\left(0, -\frac{4}{3}\right)$ and a slope of $\frac{1}{3}$. The shading goes below (and to the right of) the line. The inequality $2x + y \geq 8$ is graphed with the line that has a y-intercept of 8 and a slope of -2. The shading goes above (and to the right of) the line. The intersection of the two graphs is the area shown in the graph. A sample solution is the point (6,0).

2.

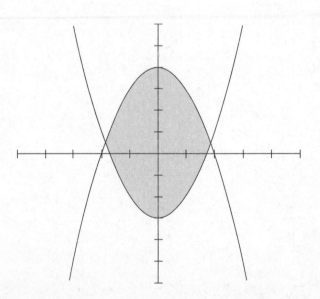

The two curves are each parabolas, one opening upward and one opening downward. The parabola opening upward corresponds to the inequality $y \geq x^2 - 3$, where everything above (inside) the curve is shaded. The parabola opening downward corresponds to the inequality $y \geq 4 - x^2$, where everything under (inside) the curve is shaded. They share solutions in the region between them. A sample solution is the point (1,1).

3.

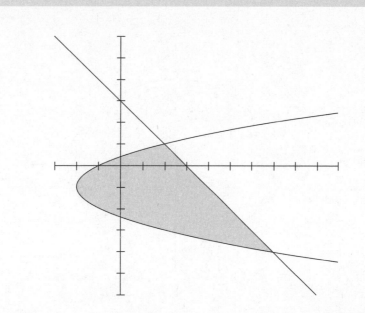

The inequality $x \geq y^2 + 2y - 1$ is represented by the parabola with the shading to the right (and inside) the parabola. The inequality $y \leq 3 - x$ is represented by the line and shading on the left (and under) the line. Their intersection is shown in the graph. A representative of the common solution is the point $(-1, -1)$.

4.

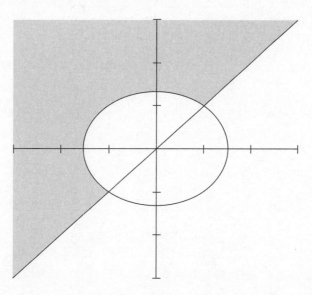

The inequality $3x^2 + 4y^2 \geq 7$ is represented by the region outside the ellipse. The inequality $y \leq x$ is represented by the region above (and to the left of) the line. Their shared solution is shown in the graph. A representative point in their common solution is the point $(-5, 5)$.

5.

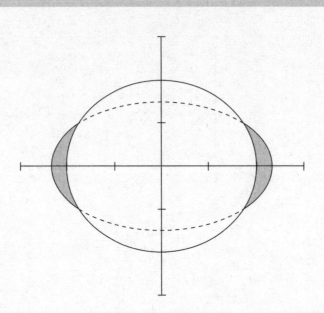

The inequality $2x^2 + 5y^2 \le 11$ is represented by the inside of the ellipse. The inequality $x^2 + y^2 \ge 4$ is represented by the region outside the circle. They share all the points in the small crescents at either end of the figure, as shown in the graph.

Chapter 10
Systems of Linear Equations with Three or More Variables

S ystems of linear equations with two variables can be solved in many different ways. In Chapter 9, the addition method, substitution, and Cramer's Rule are discussed for solving those types of systems. In this chapter, you will find the addition method again, which can be used to solve systems of three or more linear equations. After the value of a variable is found by the addition method, the values of the remaining variables are found by substitution. Another method that can be used is to incorporate matrices into the process of solving these systems. If you have a graphing calculator, solving large systems of linear equations can be done quite easily with matrices.

Addition Method

Solving systems of linear equations using the addition method (also called elimination or linear combinations) involves choosing a variable to eliminate from the equations and then using addition of the equations to accomplish that. This process can be repeated until the value of one of the variables is determined, and then "back substitution" provides the rest of the solutions. Careful selection of the variable to be eliminated or "added out" can make the computations easier. The best choice for the variable to get rid of is when at least one of the coefficients of the variable is a ± 1. This is useful because then multiples of that variable don't involve fractions when they're added to the other equations to eliminate the variable.

Example Problems
These problems show the answers and solutions.

1. Solve the system using the addition method:
$$3x + 2y + z = 12$$
$$x - 5y + 2z = 13.$$
$$2x - y + 4z = 17$$

   answer: $x = 4$, $y = -1$, $z = 2$ or $(4, -1, 2)$

Both the x and z variables have equations where the coefficients is 1. The coefficient of -1 on the y variable could be used also. For example, choose to eliminate the x's, because the coefficients of the x terms in the first and third equations are slightly smaller than some of those on the other variables. Multiply each term in the second equation by -3 and add this new equation to the first equation:

$$3x + 2y + z \quad = 12$$
$$\underline{-3x + 15y - 6z = -39}$$
$$17y - 5z = -27$$

Now, going back to the original three equations, multiply each term in the second equation by -2 and add it to the third equation.

$$-2x + 10y - 4z = -26$$
$$\underline{2x - y \quad + 4z = 17}$$
$$9y \qquad = -9$$

There's an added bonus, because the z's were eliminated, also. Solving for y, $y = -1$. Substitute that into the equation that was obtained from the first addition of equations, and $17y - 5z = -27$, $17(-1) - 5z = -27$, $-17 - 5z = -27$, $-5z = -10$, $z = 2$. Now substitute the $y = -1$ and $z = 2$ back into the first *original* equation and $3x + 2(-1) + 2 = 12$, $3x = 12$, $x = 4$.

2. Solve the system of equations using the addition method: $\begin{aligned} 2x - 3y + z &= 4 \\ 3x + 2y + z &= -2. \\ 4x - 5y + z &= 4 \end{aligned}$

    answer: $x = -1$, $y = -1$, $z = 3$ or $(-1, -1, 3)$

The best choice of variable to eliminate is the z, since the multiplications will only involve -1. First, multiply each term in the top equation by -1, and add the result to the second equation:

$$-2x + 3y - z = -4$$
$$\underline{3x + 2y + z = -2}$$
$$x + 5y \quad = -6$$

Now multiply each term in the first equation by -1 and add the result to the third equation.

$$-2x + 3y - z = -4$$
$$\underline{4x - 5y + z = 4}$$
$$2x - 2y \quad = 0$$

The two results of the addition each have two variables. This is a new system of two equations with two unknown variables that can be solved by the addition method.

$$x + 5y = -6$$
$$2x - 2y = 0$$

Multiply each term in the second equation by $-\frac{1}{2}$ and add the result to the first equation.

$$x + 5y = -6$$
$$\underline{-x + y = 0}$$
$$6y = -6$$

Solving for y, $y = -1$. Substituting back into the top equation of the new system, $x + 5(-1) = -6$, $x - 5 = -6$, $x = -1$. Now the values $x = -1$ and $y = -1$ can be substituted back into the original equation $2x - 3y + z = 4$. It doesn't matter which of the original equations you choose; they all should work: $2(-1) - 3(-1) + z = 4$; $-2 + 3 + z = 4$; $1 + z = 4$; $z = 3$.

$$3x - y + 2z + 4w - t = 12$$
$$2x + 2y - z + w + 2t = 4$$

3. Solve the system of equations using the addition method: $x + 3y + 3z - w + t = 14$.
$$4x - 2y + z + 3w - 3t = 5$$
$$x - 3y + 2z - 5w - 2t = 7$$

answer: $x = 2$, $y = -1$, $z = 4$, $w = 0$, $t = 3$ or $(2, -1, 4, 0, 3)$

Each variable has a term in at least one of the equations that has a coefficient of 1, so any would work nicely in this first elimination of a variable. For example, choose to get rid of the t's. Multiplying the terms in the first equation by 2 and adding the result to the second equation gives the following:

$$6x - 2y + 4z + 8w - 2t = 24$$
$$\underline{2x + 2y - z + w + 2t = 4}$$
$$8x + 3z + 9w = 28$$

The t can be eliminated without any changing of the equations when the first and third equations are added together.

$$3x - y + 2z + 4w - t = 12$$
$$\underline{x + 3y + 3z - w + t = 14}$$
$$4x + 2y + 5z + 3w = 26$$

Multiplying each term in the third equation by 3 and adding the results to the fourth equation gives you the following.

$$3x + 9y + 9z - 3w + 3t = 42$$
$$\underline{4x - 2y + z + 3w - 3t = 5}$$
$$7x + 7y + 10z = 47$$

And, finally, adding the second and fifth equations together you get the following.

$$2x + 2y - z + w + 2t = 4$$
$$\underline{x - 3y + 2z - 5w - 2t = 7}$$
$$3x - y + z - 4w = 11$$

Putting the four equations that resulted from the additions into a single system, there's now a system of four equations with four variables. A variable will be eliminated from that system by the addition method.

$$8x + 3z + 9w = 28$$
$$4x + 2y + 5z + 3w = 26$$
$$7x + 7y + 10z = 47$$
$$3x - y + z - 4w = 11$$

A good candidate for a variable to get rid of is one that's already missing in one of the equations. The variable y is my choice, because the last equation has a term of y with a coefficient of -1. Multiply each term in the last equation by 2 and add the results to the second equation.

$$4x + 2y + 5z + 3w = 26$$
$$\underline{6x - 2y + 2z - 8w = 22}$$
$$10x + 7z - 5w = 48$$

Now multiply each term in the last equation by 7 and add the results to the third equation.

$$7x + 7y + 10z = 47$$
$$\underline{21x - 7y + 7z - 28w = 77}$$
$$28x + 17z - 28w = 124$$

Make a system of equations using the first equation and the two addition results.

$$8x + 3z + 9w = 28$$
$$10x + 7z - 5w = 48$$
$$28x + 17z - 28w = 124$$

In this system, there is no really good choice of variable to eliminate. None of the coefficients is 1. Choose to get rid of the z's only because the coefficient 3 is the smallest number there, and the multiplications will be smaller.

Multiply each term in the first equation by -7 and each term in the second equation by 3. Then add the two equations together.

$$-56x - 21z - 63w = -196$$
$$\underline{30x + 21z - 15w = 144}$$
$$-26x - 78w = -52$$

This time, multiply each term in the first equation by -17 and each term in the third equation by 3. Then add the two equations together.

$$-136x - 51z - 153w = -476$$
$$\underline{84x + 51z - 84w = 372}$$
$$-52x - 237w = -104$$

The two resulting equations form a system of equations with two variables.

$$-26x - 78w = -52$$
$$-52x - 237w = -104$$

Multiply each term in the first equation by -2 and add the two equations together.

$$52x + 156w = 104$$
$$\underline{-52x - 237w = -104}$$
$$-81w = 0$$

Solving for w, $w = 0$. Substitute that into $52x + 156w = 104$, and you get $x = 2$. Put the values for w and x into $84x + 51z - 84w = 372$, and you get $84(2) + 51z - 84(0) = 372$, $168 + 51x = 372$, $51z = 204$, or $z = 4$. Then putting those three values into $7x + 7y + 10z = 47$, $7(2) + 7y + 10(4) = 47$, $14 + 7y + 40 = 47$, $7y = -7$, $y = -1$. Finally, with $w = 0$, $x = 2$, $z = 4$, $y = -1$, put those into $3x - y + 2z + 4w - t = 12$ and $3(2) - (-1) + 2(4) + 4(0) - t = 12$, $6 + 1 + 8 + 0 - t = 12$, $-t = -3$, $t = 3$.

4. Solve the system of equations.

$$3x - 2y + z = 3$$
$$x + y + 2z = 5$$
$$x - 4y - 3z = -7$$

Multiply the terms in the second equation by -3 and add them to the terms in the first equation. Then multiply the terms in the second equation by -1 and add them to the third equation.

$$
\begin{array}{rl}
3x - 2y + z = 3 \\
\underline{-3x - 3y - 6z = -15} \\
-5y - 5z = -12
\end{array}
\qquad
\begin{array}{rl}
x - 4y - 3z = -7 \\
\underline{-x - y - 2z = -5} \\
-5y - 5z = -12
\end{array}
$$

The two new equations are exactly the same. This happens when there is more than one solution to the system of equations. You write the multiple solutions by choosing a parameter, t, to represent one of the variables. Then the other two variables are written in terms of that parameter. For this problem, let t represent the variable z. Then use one of the new equations to solve for y in terms of t.

$$
\begin{aligned}
-5y - 5t &= -12 \\
-5t &= 5y - 12 \\
12 - 5t &= 5y \\
\frac{12 - 5t}{5} &= y
\end{aligned}
$$

Now replace the y and z in the second original equation with their equivalents in terms of t. Solve for x in terms of t.

$$
\begin{aligned}
x + y + 2z &= 5 \\
x + \frac{12 - 5t}{5} + 2t &= 5 \\
x &= 5 - \frac{12 - 5t}{5} - 2t \\
&= \frac{25}{5} - \frac{12 - 5t}{5} - \frac{10t}{5} \\
&= \frac{25 - 12 + 5t - 10t}{5} = \frac{13 - 5t}{5}
\end{aligned}
$$

Now you write the solution of the system as: $\left(\dfrac{13 - 5t}{5}, \dfrac{12 - 5t}{5}, t \right)$.

By substituting in any real numbers for t, you get the numerical solutions. For example, if you let $t = \dfrac{2}{5}$, then

$$
\left(\frac{13 - 5\left(\frac{2}{5}\right)}{5}, \frac{12 - 5\left(\frac{2}{5}\right)}{5}, \left(\frac{2}{5}\right) \right) = \left(\frac{13 - 2}{5}, \frac{12 - 2}{5}, \frac{2}{5} \right) = \left(\frac{11}{5}, 2, \frac{2}{5} \right)
$$

Work Problems

Use these problems to give yourself additional practice. Solve each system using the addition method.

1.
$$
\begin{aligned}
5a + b - 3c &= 20 \\
4a - b + 2c &= 1 \\
3a + 2b - c &= 11
\end{aligned}
$$

$$x + 6y - z = 13$$

2. $$2x - y - 4z = 6$$
    $$3x + 4y - 2z = 8$$

$$x - 3y + z = 3$$

3. $$4x + 5y + 2z = 16$$
    $$3x - 2y - z = 17$$

$$x + 2y - 3z + 4w = 12$$

4. $$2x - y + z + 2w = 11$$
    $$3x - 2y + 5z - w = 10$$
    $$5x + y - z + w = 8$$

$$2a + b - 3c + 4d = 11$$

5. $$a - 3b - c + d = -7$$
    $$3a + 5b - 2c + 6d = 17$$
    $$4a - 3b - 2c + 5d = -9$$

6. Solve the system of equations.

$$x + y + z = 7$$
$$2x - y + z = 9$$
$$3x + 6y + 4z = 26$$

Worked Solutions

1. **$a = 2, b = 1, c = -3$ or $(2,1,-3)$**   Add the first and second equations together to get rid of the terms with the variable b in them.

$$
\begin{array}{r}
5a + b - 3c = 20 \\
4a - b + 2c = 1 \\
\hline
9a \quad\quad - c = 21
\end{array}
$$

Multiply each term in the second equation by 2 and add the results to the third equation.

$$
\begin{array}{r}
8a - 2b + 4c = 2 \\
3a + 2b - c = 11 \\
\hline
11a \quad\quad + 3c = 13
\end{array}
$$

The new system has the two equations resulting from the addition. Multiply each term in the first equation by 3 and add the result to the second equation.

$$
\begin{array}{l}
9a - c = 21 \\
11a + 3c = 13
\end{array}
\text{ becomes }
\begin{array}{r}
27a \quad - 3c = 63 \\
11a \quad + 3c = 13 \\
\hline
38a \quad\quad = 76
\end{array}
$$

Divide each side of the equation $38a = 76$ by 38 to get $a = 2$. Substitute 2 for a in the equation $11a + 3c = 13$ to get $11(2) + 3c = 13$, $22 + 3c = 13$, $3c = -9$, $c = -3$. Substitute the values for a and c into the original equation $5a + b - 3c = 20$ to get $5(2) + b - 3(-3) = 20$, $10 + b + 9 = 20$, $b = 1$.

2. **$x = -2$, $y = 2$, $z = -3$ or $(-2,2,-3)$**   First multiply each term in the second equation by 6 and add the results to the first equation.

$$
\begin{aligned}
x + 6y - z &= 13 \\
12x - 6y - 24z &= 36 \\
\hline
13x \quad\quad - 25z &= 49
\end{aligned}
$$

Now multiply each term in the second equation by 4 and add the results to the third equation.

$$
\begin{aligned}
8x - 4y - 16z &= 24 \\
3x + 4y - 2z &= 8 \\
\hline
11x \quad\quad - 18z &= 32
\end{aligned}
$$

The new system consists of the two equations resulting from the addition. Multiply each term in the first equation by 11 and each term in the second equation by -13. Then add them together.

$$
\begin{matrix}
13x - 25z = 49 \\
11x - 18z = 32
\end{matrix}
\text{ becomes }
\begin{aligned}
143x - 275z &= 539 \\
-143x + 234z &= -416 \\
\hline
-41z &= 123
\end{aligned}
$$

From this, $z = -3$. Substitute this back into $13x - 25z = 49$ to get $13x - 25(-3) = 49$ or $13x + 75 = 49$, $13x = -26$, $x = -2$. Substitute the values for x and z back into the original equation $x + 6y - z = 13$, and you get $-2 + 6y - (-3) = 13$, $6y + 1 = 13$, $6y = 12$, $y = 2$.

3. **$x = 5$, $y = 0$, $z = -2$ or $(5,0,-2)$**   Multiply each term in the first equation by -4 and add the results to the second equation.

$$
\begin{aligned}
-4x + 12y - 4z &= -12 \\
4x + 5y + 2z &= 16 \\
\hline
17y - 2z &= 4
\end{aligned}
$$

Now multiply each term in the first equation by -3 and add the results to the third equation.

$$
\begin{aligned}
-3x + 9y - 3z &= -9 \\
3x - 2y - z &= 17 \\
\hline
7y - 4z &= 8
\end{aligned}
$$

Take the two resulting equations and multiply the terms in the first by -2 before adding them together.

$$
\begin{aligned}
-34y + 4z &= -8 \\
7y - 4z &= 8 \\
\hline
-27y &= 0
\end{aligned}
\text{, which means that } y = 0.
$$

Substituting that back into $7y - 4z = 8$, $0 - 4z = 8$ or $z = -2$.

Replace the y and z in the original equation $x - 3y + z = 3$ to get $x - 3(0) + (-2) = 3$, $x - 2 = 3$, $x = 5$.

4. **$x = 1$, $y = 2$, $z = 3$, $w = 4$ or (1,2,3,4)** First multiply each term in the first equation by -2 and add the results to the second equation. Then multiply each term in the first equation by -3 and add the results to the third equation. Finally multiply each term in the first equation by -5 and add them to the fourth equation. The resulting equations will not have the x variable in them.

$$
\begin{array}{r}
-2x - 4y + 6z - 8w = -24 \\
2x - y + z + 2w = 11 \\
\hline
-5y + 7z - 6w = -13
\end{array}
\qquad
\begin{array}{r}
-3x - 6y + 9z - 12w = -36 \\
3x - 2y + 5z - w = 10 \\
\hline
-8y + 14z - 13w = -26
\end{array}
\qquad
\begin{array}{r}
-5x - 10y + 15z - 20w = -60 \\
5x + y - z + w = 8 \\
\hline
-9y + 14z - 19w = -52
\end{array}
$$

The new system has three variables.

$$
\begin{array}{r}
-5y + 7z - 6w = -13 \\
-8y + 14z - 13w = -26 \\
-9y + 14z - 19w = -52
\end{array}
$$

Multiply the first equation by -2 and add the resulting equation to both the second and third equations.

$$
\begin{array}{r}
10y - 14z + 12w = 26 \\
-8y + 14z - 13w = -26 \\
\hline
2y - w = 0
\end{array}
\qquad
\begin{array}{r}
10y - 14z + 12w = 26 \\
-9y + 14z - 19w = -52 \\
\hline
y - 7w = -26
\end{array}
$$

The new system has two variables. Multiply each term in the second equation by -2 and add the results to the first equation.

$$
\begin{array}{r}
2y - w = 0 \\
y - 7w = -26
\end{array}
\qquad
\begin{array}{r}
2y - w = 0 \\
-2y + 14w = 52 \\
\hline
13w = 52
\end{array}
\qquad \text{So } w = 4.
$$

Substitute 4 for the w in the top of the two previous equations, and $2y - 4 = 0$, $y = 2$. Substitute these values for w and y into the equation $-8y + 14z - 13w = -26$ to get $-8(2) + 14z - 13(4) = -26$, $-16 + 14z - 52 = -26$, $14z = 42$, $z = 3$. Now replace the y, z, and w in the original equation $x + 2y - 3z + 4w = 12$ to get $x + 2(2) - 3(3) + 4(4) = 12$; $x + 4 - 9 + 16 = 12$; $x + 11 = 12$; $x = 1$.

5. **$a = -2$, $b = 3$, $c = -4$, $d = 0$ or (−2,3,−4,0)** Multiply each term in the second equation by -2 and add the results to the first equation. Then multiply each term in the second equation by -3 and add the results to the third equation. Finally, multiply each term in the second equation by -4 and add the results to the fourth equation.

$$
\begin{array}{r}
2a + b - 3c + 4d = 11 \\
-2a + 6b + 2c - 2d = 14 \\
\hline
7b - c + 2d = 25
\end{array}
\qquad
\begin{array}{r}
-3a + 9b + 3c - 3d = 21 \\
3a + 5b - 2c + 6d = 17 \\
\hline
14b + c + 3d = 38
\end{array}
\qquad
\begin{array}{r}
-4a + 12b + 4c - 4d = 28 \\
4a - 3b - 2c + 5d = -9 \\
\hline
9b + 2c + d = 19
\end{array}
$$

The new system has three equations and three variables.

$$
\begin{array}{r}
7b - c + 2d = 25 \\
14b + c + 3d = 38 \\
9b + 2c + d = 19
\end{array}
$$

Add the first and second equations together. Multiply the first equation by 2 and add the results to the third equation.

$$
\begin{array}{r}
7b - c + 2d = 25 \\
14b + c + 3d = 38 \\
\hline
21b + 5d = 63
\end{array}
\qquad
\begin{array}{r}
14b - 2c + 4d = 50 \\
9b + 2c + d = 19 \\
\hline
23b + 5d = 69
\end{array}
$$

Multiply the first of these results by -1 and add the resulting equation to the second.

$$
\begin{aligned}
-21b - 5d &= -63 \\
23b + 5d &= 69 \\
\hline
2b \qquad &= 6
\end{aligned}
\qquad \text{So } b = 3.
$$

Substitute this back into the second equation to get $23(3) + 5d = 69$, $69 + 5d = 69$, $5d = 0$, $d = 0$.

Put the values for b and d into the equation $7b - c + 2d = 25$ to get $7(3) - c + 2(0) = 25$, $21 - c + 0 = 25$, $-c = 4$, $c = -4$.

Last, put these three values back into the original equation $2a + b - 3c + 4d = 11$, $2a + 3 - 3(-4) + 4(0) = 11$, $2a + 3 + 12 + 0 = 11$, $2a = -4$, $a = -2$.

6. Multiply the terms in the first equation by -2 and add them to the terms in the second equation. Then multiply the terms in the first equation by -3 and add them to the third equation.

$$
\begin{aligned}
-2x - 2y - 2z &= -14 \\
2x - y + z &= 9 \\
\hline
-3y - z &= -5
\end{aligned}
\qquad
\begin{aligned}
-3x - 3y - 3z &= -21 \\
3x + 6y + 4z &= 26 \\
\hline
3y + z &= 5
\end{aligned}
$$

The two new equations are exactly the same (multiply each term in the first new equation by -1). This happens when there is more than one solution to the system of equations. You write the multiple solutions by choosing a parameter, t, to represent one of the variables. Then the other two variables are written in terms of that parameter. For this problem, let t represent the variable y. Then use one of the new equations to solve for z in terms of t.

$$
3t + z = 5
$$
$$
z = 5 - 3t
$$

Now replace the y and z in the first original equation with their equivalents in terms of t. Solve for x.

$$
x + y + z = 7
$$
$$
x + t + 5 - 3t = 7
$$
$$
x - 2t + 5 = 7
$$
$$
x = 7 + 2t - 5 = 2 + 2t
$$

You can write the solution of the system as: $(2 + 2t, t, 5 - 3t)$. Choose any number for t, and replace t in the format for the solution. For example, if $t = 1$, then you get $(2 + 2(1), 1, 5 - 3(1)) = (4, 1, 2)$.

Solving Linear Systems Using Matrices

A matrix is a rectangular array of numbers with rows and columns. Each row has the same number of values or entries in it, and each column has the same number of values in it. A bracket around the array signifies to you that it's a matrix. What's shown here is a matrix with two rows and three columns. By convention, this is written 2×3. It's *dimension* is 2×3.

$$
\begin{bmatrix}
1 & 2 & 3 \\
0 & 4 & 5
\end{bmatrix}
$$

Matrices can be used to solve systems of linear equations by assigning all of the coefficients of the variables to one matrix, finding the *inverse* of that matrix, and multiplying that result by a matrix formed by the constants. The only catch here is that the variables are no longer apparent, so the order that numbers are written in is important. The equations that the matrices are written from should have the variables all written in the same order.

The *inverse* of a matrix is discussed in the next section. For all practical purposes, however, these problems are done with a calculator. As nice as matrices are, they can still get a bit messy as the number of variables involved increases.

Example Problems
These problems show the answers and solutions.

1. Solve the system of equations using matrices.

$$\begin{aligned} x + 6y - z &= 13 \\ 2x - y - 4z &= 6 \\ 3x + 4y - 2z &= 8 \end{aligned}$$

answer: $x = -2, y = 2, z = -3$ or $(-2, 2, -3)$

The *coefficient matrix* is $\begin{bmatrix} 1 & 6 & -1 \\ 2 & -1 & -4 \\ 3 & 4 & -2 \end{bmatrix} = A$.

In this example, the matrix is named "A," because most graphing calculators name the matrices A, B, C, and so on. Matrix A is a 3×3 matrix. It's a *square* matrix, because the number of rows and columns is the same. Only square matrices have inverses, so only systems of equations with the same number of equations as there are variables can be solved in this way. The coefficient matrix will be square. The *constant* matrix will be a *column* matrix. It'll have only one column and as many rows as there are variables. In this case, I'll name

the constant matrix B: $B = \begin{bmatrix} 13 \\ 6 \\ 8 \end{bmatrix}$. The next step is to multiply the *inverse* of matrix A times

matrix B. The result will be a column matrix like B, but this column matrix will have the values of the variables, in order, from top to bottom.

To find the inverse of matrix A, enter [A] and then enter the "reciprocal" function, usually x^{-1}. The result will be $[A]^{-1}$ on your calculator screen. Then multiply this by matrix B,

giving you $[A]^{-1} * [B] = \begin{bmatrix} -2 \\ 2 \\ -3 \end{bmatrix}$. There's your answer! $x = -2, y = 2, z = -3$.

2. Solve the system using matrices.

$$\begin{aligned} x - y + 3z &= 8 \\ 5x + z - w &= 7 \\ 3x + 2y + w &= 4 \\ y + 2z + 3w &= 11 \end{aligned}$$

answer: $x = 2, y = -3, z = 1, w = 4$.

Notice that the variables are not lined up under one another. It's very important to rewrite the system before trying to write the coefficient matrix. The missing variables will be represented by 0's.

$$\begin{aligned} x - y + 3z &= 8 \\ 5x + z - w &= 7 \\ 3x + 2y + w &= 4 \\ y + 2z + 3w &= 11 \end{aligned} \quad \text{has the coefficient matrix A} = \begin{bmatrix} 1 & -1 & 3 & 0 \\ 5 & 0 & 1 & -1 \\ 3 & 2 & 0 & 1 \\ 0 & 1 & 2 & 3 \end{bmatrix}.$$

Multiply the inverse of matrix A by the constant matrix B: $\quad B = \begin{bmatrix} 8 \\ 7 \\ 4 \\ 11 \end{bmatrix}.$

So the multiplication is $[A]^{-1} * [B] = \begin{bmatrix} 2 \\ -3 \\ 1 \\ 4 \end{bmatrix}.$ Therefore, $x = 2$, $y = -3$, $z = 1$, and $w = 4$.

Work Problem

Use these problems to give yourself additional practice. Solve each system using matrices.

1.
$$\begin{aligned} a - 5b - 6c &= 10 \\ 2a + b - 3c &= 0 \\ 4a - 6b - c &= 3 \end{aligned}$$

2.
$$\begin{aligned} 2x - 4y + 3z &= 4 \\ x - 2y - 2z &= 9 \\ 3x + 6y - 2z &= 13 \end{aligned}$$

3.
$$\begin{aligned} x - y + 3z + w &= 14 \\ 2x + 9z - w &= 5 \\ 2y + 3z + 4w &= 19 \\ 3x - 6z - w &= 4 \end{aligned}$$

4.
$$\begin{aligned} 3x - y + 2z - w &= 2 \\ 5x + 3y + 2w &= 18 \\ 4x - 5y - 7z + w &= 12 \\ 9z + 4w &= 16 \end{aligned}$$

5.
$$\begin{aligned} 2x + 3y - 5z &= 1 \\ 6x - 9y + 10z &= 2 \\ 4x - 6y + 15z &= 3 \end{aligned}$$

Worked Solutions

1. $a = -1$, $b = -1$, $c = -1$ **or** $(-1, -1, -1)$

The coefficient matrix is $A = \begin{bmatrix} 1 & -5 & -6 \\ 2 & 1 & -3 \\ 4 & -6 & -1 \end{bmatrix}$, and the constant matrix is $B = \begin{bmatrix} 10 \\ 0 \\ 3 \end{bmatrix}.$

Multiplying the inverse of matrix A by matrix B, $[A]^{-1} * [B] = \begin{bmatrix} -1 \\ -1 \\ -1 \end{bmatrix}.$

2. $x = 4$, $y = -\frac{1}{2}$, $z = -2$ or $\left(4, -\frac{1}{2}, -2\right)$

The coefficient matrix is A = $\begin{bmatrix} 2 & -4 & 3 \\ 1 & -2 & -2 \\ 3 & 6 & -2 \end{bmatrix}$, and the constant matrix is B = $\begin{bmatrix} 4 \\ 9 \\ 13 \end{bmatrix}$.

Multiplying the inverse of matrix A by matrix B, $[A]^{-1}*[B] = \begin{bmatrix} 4 \\ -\frac{1}{2} \\ -2 \end{bmatrix}$.

3. $x = 4$, $y = -3$, $z = \frac{1}{3}$, $w = 6$ or $\left(4, -3, \frac{1}{3}, 6\right)$

The coefficient matrix is A = $\begin{bmatrix} 1 & -1 & 3 & 1 \\ 2 & 0 & 9 & -1 \\ 0 & 2 & 3 & 4 \\ 3 & 0 & -6 & -1 \end{bmatrix}$, and the constant matrix is B = $\begin{bmatrix} 14 \\ 5 \\ 19 \\ 4 \end{bmatrix}$.

Multiplying the inverse of matrix A by matrix B, $[A]^{-1}*[B] = \begin{bmatrix} 4 \\ -3 \\ \frac{1}{3} \\ 6 \end{bmatrix}$.

4. $x = 2$, $y = 0$, $z = 0$, $w = 4$ or $(2, 0, 0, 4)$

The coefficient matrix is A = $\begin{bmatrix} 3 & -1 & 2 & -1 \\ 5 & 3 & 0 & 2 \\ 4 & -5 & -7 & 1 \\ 0 & 0 & 9 & 4 \end{bmatrix}$, and the constant matrix is B = $\begin{bmatrix} 2 \\ 18 \\ 12 \\ 16 \end{bmatrix}$.

Multiplying the inverse of matrix A by matrix B, $[A]^{-1}*[B] = \begin{bmatrix} 2 \\ 0 \\ 0 \\ 4 \end{bmatrix}$.

5. $x = \frac{1}{2}$, $y = \frac{1}{3}$, $z = \frac{1}{5}$ or $\left(\frac{1}{2}, \frac{1}{3}, \frac{1}{5}\right)$

The coefficient matrix is A = $\begin{bmatrix} 2 & 3 & -5 \\ 6 & -9 & 10 \\ 4 & -6 & 15 \end{bmatrix}$, and the constant matrix is B = $\begin{bmatrix} 1 \\ 2 \\ 3 \end{bmatrix}$.

Multiplying the inverse of matrix A by matrix B, $[A]^{-1}*[B] = \begin{bmatrix} \frac{1}{2} \\ \frac{1}{3} \\ \frac{1}{5} \end{bmatrix}$.

Operations on Matrices

One use of matrices is discussed in the previous section where systems of linear equations can be solved readily by performing operations on matrices with a graphing calculator. Many operations on matrices are much simpler than those done with the graphing calculator. These simpler operations can be done with pencil and paper. The many applications of matrices make use of the operations, and you will have a chance to become familiar with them here. Matrices are a means of organizing large groups of data, making them easy to read and comprehend. This form allows for the manipulations needed. Many calculator and computer spreadsheets can quickly and easily do matrix operations.

The operations shown in this section are addition, subtraction, scalar multiplication, and matrix multiplication. You will also see a method for finding the inverse of a 2 × 2 matrix.

Addition and Subtraction of Matrices

Matrices can be added and subtracted only if they're the same size (*dimension*). They have to have the same number of rows and the same number of columns to be added and subtracted. To add or subtract matrices, you add or subtract, individually, the corresponding entries. You add the first matrices' entry in the first row, second column to the other matrices' entry in the first row, second column.

Example Problems

These problems show the answers and solutions.

In each problem, use the matrices $A = \begin{bmatrix} 2 & 3 & -4 & 1 \\ -5 & 6 & 0 & 2 \end{bmatrix}$ and $B = \begin{bmatrix} -1 & 3 & 6 & 4 \\ -2 & -3 & -4 & -5 \end{bmatrix}$.

1. Find A + B.

    answer: $\begin{bmatrix} 1 & 6 & 2 & 5 \\ -7 & 3 & -4 & -3 \end{bmatrix}$

    Add the entries that are in the same positions together.

    $$A + B = \begin{bmatrix} 2 & 3 & -4 & 1 \\ -5 & 6 & 0 & 2 \end{bmatrix} + \begin{bmatrix} -1 & 3 & 6 & 4 \\ -2 & -3 & -4 & -5 \end{bmatrix} =$$

    $$\begin{bmatrix} 2+(-1) & 3+3 & -4+6 & 1+4 \\ -5+(-2) & 6+(-3) & 0-(-4) & 2+(-5) \end{bmatrix}$$

2. Find A − B.

    answer: $\begin{bmatrix} 3 & 0 & -10 & -3 \\ -3 & 9 & 4 & 7 \end{bmatrix}$

    $$A - B = \begin{bmatrix} 2 & 3 & -4 & 1 \\ -5 & 6 & 0 & 2 \end{bmatrix} - \begin{bmatrix} -1 & 3 & 6 & 4 \\ -2 & -3 & -4 & -5 \end{bmatrix} =$$

    $$\begin{bmatrix} 2-(-1) & 3-3 & -4-6 & 1-4 \\ -5-(-2) & 6-(-3) & 0-(-4) & 2-(-5) \end{bmatrix}$$

3. Find B − A.

    answer: $\begin{bmatrix} -3 & 0 & 10 & 3 \\ 3 & -9 & -4 & -7 \end{bmatrix}$

    Changing the order of the subtraction changes the final answer. Each entry in the answer to B − A is the opposite of that in A − B.

Scalar Multiplication

Scalar multiplication of matrices means to multiply each entry in the matrix by the same number. It doesn't matter what size the matrix is. Scalar multiplication is indicated by writing the multiplying number in front of the matrix.

Example Problems

These problems show the answers and solutions.

In each problem, use the matrix $A = \begin{bmatrix} 2 & 3 & -4 & 1 \\ -5 & 6 & 0 & 2 \end{bmatrix}$.

1. Find 5A.

   answer: $\begin{bmatrix} 10 & 15 & -20 & 5 \\ -25 & 30 & 0 & 10 \end{bmatrix}$

   Multiply each entry by 5.

$$5 \begin{bmatrix} 2 & 3 & -4 & 1 \\ -5 & 6 & 0 & 2 \end{bmatrix} = \begin{bmatrix} 5\cdot2 & 5\cdot3 & 5(-4) & 5\cdot1 \\ 5(-5) & 5\cdot6 & 5\cdot0 & 5\cdot2 \end{bmatrix}$$

2. Find −2.3A.

   answer: $\begin{bmatrix} -4.6 & -6.9 & 9.2 & -2.3 \\ 11.5 & -13.8 & 0 & -4.6 \end{bmatrix}$

   Multiply each entry by −2.3.

$$-2.3 \begin{bmatrix} 2 & 3 & -4 & 1 \\ -5 & 6 & 0 & 2 \end{bmatrix} = \begin{bmatrix} -2.3\cdot2 & -2.3\cdot3 & -2.3\cdot(-4) & -2.3\cdot1 \\ -2.3(-5) & -2.3\cdot6 & -2.3\cdot0 & -2.3\cdot2 \end{bmatrix}$$

Matrix Multiplication

Matrix multiplication is quite different from scalar multiplication. Multiplying two matrices together takes careful alignment of the elements and accurate computation. Also, two matrices can be multiplied together only if the number of *columns* in the first matrix is the same as the number of *rows* in the second matrix. The result of this multiplication is another matrix that has the number of *rows* of the first matrix and the number of *columns* of the second matrix. The best way to explain how this multiplication works is to show you examples.

Example Problems

These problems show the answers and the solutions.

1. Multiply matrix $A = \begin{bmatrix} 1 & 2 & 3 \\ 4 & 5 & 6 \end{bmatrix}$ times matrix $B = \begin{bmatrix} 7 & 8 \\ 9 & 0 \\ -1 & -2 \end{bmatrix}$.

   answer: $A*B = \begin{bmatrix} 22 & 2 \\ 67 & 20 \end{bmatrix}$

In this example, A is a 2 × 3 matrix, and B is a 3 × 2 matrix. The number of columns in A matches the number of rows in B, so they can be multiplied together **in that order**. The answer matrix is 2 × 2, which is the number of rows in A and the number of columns in B.

To get the entries that belong in that answer, think of the result of the multiplication as being the matrix C that looks like this: $C = \begin{bmatrix} p & q \\ r & t \end{bmatrix}$. The entry p is in the *first row, first column*. You get the entry by multiplying the entries of the *first row* of matrix A times the *first column* of matrix B and adding the products. So $p = 1 \cdot 7 + 2 \cdot 9 + 3(-1) = 7 + 18 - 3 = 22$. The entry q is in the *first row, second column*. You get this entry by multiplying the entries of the *first row* of matrix A times the *second column* of matrix B and adding the products. So $q = 1 \cdot 8 + 2 \cdot 0 + 3(-2) = 8 + 0 - 6 = 2$. The entry *r* is in the *second row, first column*. You get the entry by multiplying the entries of the *second row* of matrix A times the *first column* of matrix B and adding the products. So $r = 4 \cdot 7 + 5 \cdot 9 + 6(-1) = 28 + 45 - 6 = 67$. And, finally, the entry *t* is in the *second row, second column*. You get this entry by multiplying the entries of the *second row* of matrix A times the *second column* of matrix B and adding the products. So $t = 4 \cdot 8 + 5 \cdot 0 + 6(-2) = 32 + 0 - 12 = 20$.

2. Multiply matrix $A = \begin{bmatrix} 1 & 3 \\ -2 & 4 \end{bmatrix}$ times matrix $B = \begin{bmatrix} 1 & 2 & 3 \\ 5 & 6 & 7 \end{bmatrix}$.

   answer: $\begin{bmatrix} 16 & 20 & 24 \\ 18 & 20 & 22 \end{bmatrix}$

The product will consist of the following sums of multiplications:

$$\begin{bmatrix} \text{1st row A, 1st columnB} & \text{1st row A, 2nd column B} & \text{1st row A, 3rd column B} \\ \text{2nd row A, 1st column B} & \text{2nd row A, 2nd column B} & \text{2nd row A, 3rd column B} \end{bmatrix}$$

$$= \begin{bmatrix} 1 \cdot 1 + 3 \cdot 5 & 1 \cdot 2 + 3 \cdot 6 & 1 \cdot 3 + 3 \cdot 7 \\ -2 \cdot 1 + 4 \cdot 5 & -2 \cdot 2 + 4 \cdot 6 & -2 \cdot 3 + 4 \cdot 7 \end{bmatrix} = \begin{bmatrix} 16 & 20 & 24 \\ 18 & 20 & 22 \end{bmatrix}$$

Inverse of a 2 × 2 Matrix

The formula for finding the inverse of a 2×2 matrix resembles the process used in Cramer's Rule, found in Chapter 9. If A is a square matrix, then its inverse is denoted A^{-1}. The inverse is used when solving systems of linear equations. The inverse of a matrix is much like the multiplicative inverse of a number. When you multiply 3 times its multiplicative inverse $\frac{1}{3}$, you get the number 1, which is the identity for multiplication. When you multiply matrix A times its inverse A^{-1}, you get the identity matrix. This is a matrix that's square and has 1's down the diagonal from left to right and 0's elsewhere.

Consider the 2×2 matrix $A = \begin{bmatrix} a & b \\ c & d \end{bmatrix}$. Its inverse, $A^{-1} = \begin{bmatrix} \frac{d}{\Delta} & -\frac{b}{\Delta} \\ -\frac{c}{\Delta} & \frac{a}{\Delta} \end{bmatrix}$ where $\Delta = ad - bc$.

So, if $A = \begin{bmatrix} -3 & 2 \\ -7 & 5 \end{bmatrix}$, then $\Delta = -3 \cdot 5 - 2(-7) = -15 + 14 = -1$. Then $A^{-1} = \begin{bmatrix} \frac{5}{-1} & \frac{2}{-1} \\ \frac{-7}{-1} & \frac{-3}{-1} \end{bmatrix}$.

To show that this matrix is an inverse of matrix A, multiply them together:

$A * A^{-1} = \begin{bmatrix} -3 & 2 \\ -7 & 5 \end{bmatrix} * \begin{bmatrix} -5 & 2 \\ -7 & 3 \end{bmatrix} = \begin{bmatrix} 1 & 0 \\ 0 & 1 \end{bmatrix}$. Although matrix multiplication is not generally commutative (you won't get the same answer when reversing the order), in the case of a matrix and its inverse, it doesn't matter what order you multiply them in. You'll always get that identity matrix.

Work Problems

Use these problems to give yourself additional practice.

In these problems, use the matrices $A = \begin{bmatrix} 8 & 3 \\ 5 & 2 \end{bmatrix}$, $B = \begin{bmatrix} -6 & 7 \\ 4 & -5 \end{bmatrix}$, $C = \begin{bmatrix} 4 & 1 & -6 \\ 3 & 8 & 2 \end{bmatrix}$, $D = \begin{bmatrix} 9 \\ 1 \\ 7 \end{bmatrix}$, $E = \begin{bmatrix} 1 & 0 & -3 \\ 4 & 3 & 8 \end{bmatrix}$.

1. Which of the matrices can be added, subtracted, and multiplied?

2. Find C − E and A + B.

3. Find 6C.

4. Find BC.

5. Find the inverse of A.

Worked Solutions

1. **The matrices that can be added are A + B and C + E. The matrices that can be subtracted are the same as those that can be added. The matrices that can be multiplied are AB, BA, AC, AE, BC, BE, CD, and ED.** Matrices that can be added are those that are the same dimension (size). They can be added in either order, and the answer will come out the same. To subtract matrices, they have to be the same size, too, but reversing the order will result in different answers.

   Matrices that can be multiplied are those in which the number of columns in the first matrix matches the number of rows in the second matrix. Notice that one answer was AC, in that order. CA won't work.

2. $C - E = \begin{bmatrix} 3 & 1 & -3 \\ -1 & 5 & -6 \end{bmatrix}$ **and** $A + B = \begin{bmatrix} 2 & 10 \\ 9 & -3 \end{bmatrix}$ Addition and subtraction of matrices can be done only when the matrices have the same size (dimension). The corresponding entries have the operation performed upon them to form the new entries.

   $$C - E = \begin{bmatrix} 4-1 & 1-0 & -6-(-3) \\ 3-4 & 8-3 & 2-8 \end{bmatrix} \text{ and } A + B = \begin{bmatrix} 8+(-6) & 3+7 \\ 5+4 & 2-5 \end{bmatrix}$$

3. $\begin{bmatrix} 24 & 6 & -36 \\ 18 & 48 & 12 \end{bmatrix}$ Scalar multiplication is done when each entry in the matrix is multiplied by the number multiplier.

   $$6C = 6 \begin{bmatrix} 4 & 1 & -6 \\ 3 & 8 & 2 \end{bmatrix} = \begin{bmatrix} 6\cdot4 & 6\cdot1 & 6(-6) \\ 6\cdot3 & 6\cdot8 & 6\cdot2 \end{bmatrix}$$

4. $\begin{bmatrix} -3 & 50 & 50 \\ 1 & -36 & -34 \end{bmatrix}$ Matrix multiplication involves multiplying the entries in the rows of the first matrix times the entries in the columns of the second matrix and adding those products together. The resulting matrix in this case should be 2×3.

   $$\begin{bmatrix} -6\cdot4+7\cdot3 & -6\cdot1+7\cdot8 & -6(-6)+7\cdot2 \\ 4\cdot4+(-5)3 & 4\cdot1+(-5)8 & 4(-6)+(-5)2 \end{bmatrix} = \begin{bmatrix} -24+21 & -6+56 & 36+14 \\ 16+(-15) & 4+(-40) & -24+(-10) \end{bmatrix}$$

5. $\begin{bmatrix} 2 & -3 \\ -5 & 8 \end{bmatrix}$ The value of $\Delta = 8 \cdot 2 - 3 \cdot 5 = 16 - 15 = 1$, so every entry will be divided by 1.

Applications of Systems of Linear Equations

There have been several techniques mentioned earlier in this chapter and in Chapter 9 on how to solve systems of linear equations. Refer to the sections on solving systems using the *addition method*, *substitution*, *Cramer's Rule*, and *matrices*, if you need help solving the problems in this section. Choosing a method to use when solving a problem will depend both on your personal preference and the types of technology available. The practical problems cannot be solved, however, if the system is not set up correctly. The following examples should give you some insight into the types of problems that can be solved and the solution methods that work best.

Example Problems

These problems show the answers and solutions.

1. Stephanie raises chickens, ducks, and egrets. When she sold 4 chickens, 5 ducks, and 12 egrets, she earned $166. The same dealer gave her $98 for 2 chickens and 9 egrets. One other transaction earned her $200 when she sold 10 chickens, 20 ducks, and 4 egrets. How much are each of these birds worth?

    answer: Chickens are $4, ducks are $6, and egrets are $10

    Let c represent the cost of one chicken, d represent the cost of one duck, and e represent the cost of one egret. The system of equations that can be written to show the transactions is

    $$4c + 5d + 12e = 166$$
    $$2c \quad\quad + 9e = 98$$
    $$10c + 20d + 4e = 200$$

    If the addition method is used to solve the system, the best variable to eliminate is the d, because it only appears in the first and third equations. Multiply each term in the first equation by -4 and add the result to the third equation.

    $$-16c - 20d - 48e = -664$$
    $$\underline{10c + 20d + 4e \quad = 200}$$
    $$-6c \quad\quad\quad - 44e = -464$$

    Take the original second equation and multiply each term by 3. Add the results to this new equation found by addition.

    $$6c + 27e = \quad 294$$
    $$\underline{-6c - 44e = -464}$$
    $$-17e = -170$$
    $$e = 10$$

    Substituting this value for e back into the original second equation, $2c + 9(10) = 98$, $2c + 90 = 98$, $2c = 8$, $c = 4$. Now replace the c and e in the original first equation to get $4(4) + 5d + 12(10) = 166$, $16 + 5d + 120 = 166$, $5d + 136 = 166$, $5d = 30$, $d = 6$.

2. Alex invested $2000 in bond A, $1500 in bond B, and $1000 in bond C. Betty invested $5000 in bond A, $6000 in bond B, and $1000 in bond C. Cara invested $4000 in bond A, $3000 in bond B, and $8000 in bond C. Alex earned a total of $180 interest; Betty earned $390 interest; and Cara earned $780 interest. What were the interest rates of these three bonds?

    answer: Bond A was 4%, bond B was 2%, and bond C was 7%

The formula for simple interest is $I = prt$ where I is the amount of interest earned, p is the principal (amount invested), and r is the rate at which the money is earning interest, and t is the length of time, in years, that the investment is allowed to grow. In this problem, $t=1$. Let A, B, and C represent the interest rates of the three bonds and use the following system of equations.

$$2000A + 1500B + 1000C = 180$$
$$5000A + 6000B + 1000C = 390$$
$$4000A + 3000B + 8000C = 780$$

The numbers are so large in this system that it's probably more efficient to use matrices to solve the system. Let the coefficient matrix be A and the constant matrix be B.

$$A = \begin{bmatrix} 2000 & 1500 & 1000 \\ 5000 & 6000 & 1000 \\ 4000 & 3000 & 8000 \end{bmatrix} \text{ and } B = \begin{bmatrix} 180 \\ 390 \\ 780 \end{bmatrix}$$

Using a graphing calculator, multiply the inverse of matrix A times matrix B:

$$[A]^{-1} * [B] = \begin{bmatrix} .04 \\ .02 \\ .07 \end{bmatrix}.$$ These are, from top to bottom, the interest rates in decimal form.

Therefore, bond A earns 4% interest, bond B earns 2% interest, and bond C earns 7% interest.

Work Problems

Use these problems to give yourself more practice.

1. At a local bakery, an order of 2 croissants, 3 donuts, and 10 éclairs cost $28.00. A second order of 3 croissants, 2 donuts, and 4 éclairs cost $16.00. A third order of 1 croissant, 9 donuts, and 2 éclairs cost $11.00. How much does each of these items cost?

2. Tom, Dick, and Harry sell new cars and get a commission on each one they sell. Last month, Tom sold 2 sedans, 3 SUVs, and 1 sports car and earned a commission of $1,800. Dick sold 5 sedans and 3 sports cars and earned $2,500. Harry sold 1 sedan, 4 SUVs, and 2 sports cars and earned $2,400. What is the commission on these three types of vehicles?

3. Greg needs to buy large amounts of candy bars to fill gift bags. He can get cases of almond bars, butter crunch bars, caramel bars, and dark chocolate nut bars. There are different numbers of bars in the different cases, depending on the type. One purchase he made of 6 cases of almonds bars, 3 cases of butter crunch bars, and 2 cases of dark chocolate nut bars resulted in a total of 216 candy bars. A second purchase of 10 cases of almond bars, 6 cases of caramel bars, and 3 cases of dark chocolate nut bars had a total of 434 candy bars. A third purchase was of 12 cases of almond bars, 6 cases of butter crunch bars, and 4 cases of caramel bars, which had a total of 408 candy bars. His last purchase was of 10 cases of butter crunch bars, 5 cases of caramel bars, and 8 cases of dark chocolate nut bars and had a total of 480 candy bars. How many candy bars are in each type of case?

4. In a certain state, food, soft drinks, and paper products are all taxed at different rates. A purchase of $40 in food, $15 in soft drinks, and $10 in paper products had a total tax of $1.85. When $57 in food, $40 in soft drinks, and $20 in paper products was purchased, the tax was $3.97. And a purchase including $200 in food, $25 in soft drinks, and $45 in paper products had a tax of $6.40. What are the tax rates on these different items?

5. Jack has pennies, nickels, and dimes totaling $2.28. Chloe has twice as many pennies and nickels as Jack has and the same number of dimes; her total is $2.96. Carley has the same number of pennies and nickels as Jack, but she has five times as many dimes as he has; her total is $8.68. How many pennies, nickels, and dimes does Jack have?

Worked Solutions

1. **Croissants are $2, donuts are $.50, éclairs are $2.25.**   Let the price of each croissant be represented by c, the price of each donut be represented by d, and the price of each éclair be represented by e. Then the equations to solve are as follows:

$$2c + 3d + 10e = 28.00$$
$$3c + 2d + 4e = 16.00$$
$$c + 9d + 2e = 11.00$$

The system can be solved using the addition method. Multiply each term in the third equation by -2 and add the results to the first equation; then multiply each term in the third equation by -3 and add the results to the second equation. The two resulting equations form the following system.

$$-15d + 6e = 6$$
$$-25e - 2e = -17$$

Multiply each term in the second equation by 3 and add the results to the first equation. That gives you the equation $-90d = -45$. Divide each side by -90, $d = .50$. Substitute that back into either of the two equations to get $e = 2.25$. Substitute the values for d and e back into any of the original equations to get $c = 2$.

2. **Sedans are $200, SUVs are $300, sports cars are $500.**   Let x represent the amount of commission on a sedan, y represent the amount of commission on an SUV, and z represent the amount of commission on a sports car. Then the system of equations is

$$2x + 3y + z = 1800$$
$$5x \qquad + 3z = 2500$$
$$x + 4y + 2z = 2400$$

Using the addition method and the fact that the y variable isn't in the second equation, multiply the terms in the first equation by 4 and the terms in the third equation by -3 and add the two results together to get $5x - 2z = 0$. Use this equation and the original second equation in a system to solve.

$$5x - 2z = 0$$
$$5x + 3z = 2500$$

Multiply each term in the first equation by -1 and add the two equations together to get $5z = 2500$. This tells you that $z = 500$. Substitute this back into the top of the two equations to get $x = 200$. Then put these two values back into the first or third original equation to get $y = 300$.

3. **20 bars per case of almond bars, 12 bars per case of butter crunch bars, 24 bars per case of caramel bars, 30 bars per case of dark chocolate nut bars** Let a, b, c, and d represent the number of candy bars in the almond bar cases, butter crunch bar cases, caramel bar cases, and dark chocolate nut bar cases, respectively. The system of equations that can be formed is as follows.

$$
\begin{aligned}
6a + 3b + \quad\ \ 2d &= 216 \\
10a \quad\ \ + 6c + 3d &= 434 \\
12a + 6b + 4c \quad\ \ &= 408 \\
10b + 5c + 8d &= 480
\end{aligned}
$$

This can be done with the addition method, but, with four equations, it would be easier to use matrices and a graphing calculator. Let A represent the coefficient matrix and B represent the constant matrix.

$$
A = \begin{bmatrix} 6 & 3 & 0 & 2 \\ 10 & 0 & 6 & 3 \\ 12 & 6 & 4 & 0 \\ 0 & 10 & 5 & 8 \end{bmatrix} \text{ and } B = \begin{bmatrix} 216 \\ 434 \\ 408 \\ 480 \end{bmatrix}
$$

Now multiply the inverse of matrix A by matrix B: $[A]^{-1} * [B] = \begin{bmatrix} 20 \\ 12 \\ 24 \\ 30 \end{bmatrix}$. From top to bottom, these are the values of a, b, c, and d.

4. **Tax on food is 1%, tax on soft drinks is 5%, tax on paper products is 7%.** Let f represent the tax rate on food, d represent the tax rate on soft drinks, and p represent the tax rate on paper products. The system of equations is as follows:

$$
\begin{aligned}
40f + 15d + 10p &= 1.85 \\
57f + 40d + 20p &= 3.97 \\
200f + 25d + 45p &= 6.40
\end{aligned}
$$

This system of equations is best solved using matrices. None of the coefficients is a 1, which makes using the addition method much more difficult. Let A represent the coefficient matrix and B represent the constant matrix.

$$
A = \begin{bmatrix} 40 & 15 & 10 \\ 57 & 40 & 20 \\ 200 & 25 & 45 \end{bmatrix} \text{ and } B = \begin{bmatrix} 1.85 \\ 3.97 \\ 6.40 \end{bmatrix}
$$

Now multiply the inverse of matrix A by matrix B: $[A]^{-1} * [B] = \begin{bmatrix} .01 \\ .05 \\ .07 \end{bmatrix}$. From top to bottom, these are the tax rates in decimal form.

5. **8 pennies, 12 nickels, and 16 dimes** Let p represent the number of pennies that Jack has, n the number of nickels, and d the number of dimes. Multiplying each of the types of coins by their monetary value, in decimal form, the first equation would read $.01p + .05n + .10d = 2.28$. Chloe has twice as many pennies and nickels as Jack, so $2p$ and $2n$ will be used in the equation to represent the amount of money that she has: $.01(2p) + .05(2n) + .10d = 2.96$. Carley has five times as many dimes as Jack, so $5d$ will be used in the equation to represent the amount of money that she has: $.01p + .05n + .10(5d) = 8.68$. The following system has the three equations in their simplified form.

$$
\begin{aligned}
.01p + .05n + .10d &= 2.28 \\
.02p + .10n + .10d &= 2.96 \\
.01p + .05n + .50d &= 8.68
\end{aligned}
$$

The nicest way to solve this is to use matrices and a graphing calculator. Let A represent the coefficient matrix and B represent the constant matrix.

$$A = \begin{bmatrix} .01 & .05 & .10 \\ .02 & .10 & .10 \\ .01 & .05 & .50 \end{bmatrix} \text{ and } B = \begin{bmatrix} 2.28 \\ 2.96 \\ 8.68 \end{bmatrix}$$

Now multiply the inverse of matrix A by matrix B: $[A]^{-1} * [B] = \begin{bmatrix} 8 \\ 12 \\ 16 \end{bmatrix}$. From top to bottom, the values give the number of coins that Jack has.

Customized Full-Length Exam

1. Simplify the expression leaving no negative exponents: $\left(\dfrac{xy^2}{x^{-1}y^{-2}}\right)^2 \cdot \left(\dfrac{x^3}{y^{-2}}\right)^{-1}$.

   Answer: xy^6

If you answered **correctly**, go to problem 4.
If you answered **incorrectly**, go to problem 2.

2. Simplify the expression leaving no negative exponents: $\left(\dfrac{a^2b}{cd}\right)^2 \cdot \dfrac{c^2d}{ab^2}$.

   Answer: $\dfrac{a^3}{d}$

If you answered **correctly**, go to problem 4.
If you answered **incorrectly**, go to problem 3.

3. Simplify the expression leaving no negative exponents: $\dfrac{x^2\left(x^3\right)^4}{x^5}$.

   Answer: x^9

If you answered **correctly,** go to problem 4.
If you answered **incorrectly**, review "Rules for Exponents" on page 24.

4. Simplify by combining like terms: $3z^2 - 2z^2 + z + 5 - z^2$.

   Answer: $z + 5$

If you answered **correctly**, go to problem 6.
If you answered **incorrectly**, go to problem 5.

5. Simplify by combining like terms: $2xyz + 3xy + 4xyz + x - xy$.

   Answer: $6xyz + 2xy + x$

If you answered **correctly**, go to problem 6.
If you answered **incorrectly**, review "Adding and Subtracting Polynomials" on page 26.

6. Find the simplified product: $(6x + 5)(x + 9)$.

   Answer: $6x^2 + 59x + 45$

If you answered **correctly**, go to problem 8.
If you answered **incorrectly**, go to problem 7.

7. Find the simplified product: $(xy + 3)(3xy - 2)$.

   Answer: $3x^2y^2 + 7xy - 6$

If you answered **correctly**, go to problem 8.
If you answered **incorrectly**, review "Multiplying Polynomials" on page 28.

8. Find the simplified product: $(9 - m)(9 + m)$.

    Answer: $81 - m^2$

If you answered **correctly**, go to problem 10.
If you answered **incorrectly**, go to problem 9.

9. Find the simplified product of $(t + 3)^2$.

    Answer: $t^2 + 6t + 9$

If you answered **correctly**, go to problem 10.
If you answered **incorrectly**, review "Special Products" on page 30.

10. Find the simplified product of $(y - 1)^3$.

    Answer: $y^3 - 3y^2 + 3y - 1$

If you answered **correctly**, go to problem 12.
If you answered **incorrectly**, go to problem 11.

11. Find the simplified product of $(r + 5)(r^2 - 5r + 25)$.

    Answer: $r^3 + 125$

If you answered **correctly**, go to problem 12.
If you answered **incorrectly**, review "Special Products" on page 30.

12. Factor the expression completely: $12x^{-3}y^{-2} + 16x^{-2}y^{-4}$.

    Answer: $4x^{-3}y^{-4}(3y^2 + 4x)$

If you answered **correctly**, go to problem 14.
If you answered **incorrectly**, go to problem 13.

13. Factor the expression completely: $5a^2b^3c^4 - 15ab^4c^8 + 20a^3b^3c^3$.

    Answer: $5ab^3c^3(ac - 3bc^5 + 4a^2)$

If you answered **correctly**, go to problem 14.
If you answered **incorrectly**, review "Greatest Common Factors and Factoring Binomials" on page 41.

14. Factor the expression completely: $a^6 + 27$.

    Answer: $(a^2 + 3)(a^4 - 3a^2 + 9)$

If you answered **correctly**, go to problem 16.
If you answered **incorrectly**, go to problem 15.

15. Factor the expression completely: $100 - 36t^2$.

    Answer: $4(5 - 3t)(5 + 3t)$

If you answered **correctly**, go to problem 16.
If you answered **incorrectly**, review "Factoring Binomials" on page 44.

16. Factor the expression completely: $12m^4 + 13m^2 - 14$.

    Answer: $(3m^2 - 2)(4m^2 + 7)$

If you answered **correctly**, go to problem 18.
If you answered **incorrectly**, go to problem 17.

17. Factor the expression completely: $12x^2 - 13x - 14$.

    Answer: $(4x - 7)(3x + 2)$

If you answered **correctly**, go to problem 18.
If you answered **incorrectly**, review "Factoring Trinomials" on page 46.

18. Factor the expression completely: $8a^3x^2 - 32a^3 + x^2 - 4$.

    Answer: $(2a + 1)(4a^2 - 2a + 1)(x - 2)(x + 2)$

If you answered **correctly**, go to problem 19.
If you answered **incorrectly**, review "Factoring by Grouping" on page 50.

19. Solve $12x^2 - 13x - 14 = 0$.

    Answer: $x = \dfrac{7}{4}$, $x = -\dfrac{2}{3}$

If you answered **correctly**, go to problem 21.
If you answered **incorrectly**, go to problem 20.

20. Solve $x^2 - 5x - 36 = 0$.

    Answer: $x = 9$, $x = -4$

If you answered **correctly**, go to problem 21.
If you answered **incorrectly**, review "Solving Quadratic Equations" on page 52.

21. Solve $3x^2 + 5x - 3 = 0$.

    Answer: $x = \dfrac{-5 \pm \sqrt{61}}{6}$

If you answered **correctly**, go to problem 22.
If you answered **incorrectly**, review "Solving Quadratic Equations with the Quadratic Formula" on page 55.

22. Solve $y^{-4} + y^{-2} - 2 = 0$.

    Answer: $y = \pm 1$

If you answered **correctly**, go to problem 24.
If you answered **incorrectly**, go to problem 23.

23. Solve $z^6 - 9z^3 + 8 = 0$ if z is a real number.

    Answer: $z = 1$, $z = 2$

If you answered **correctly**, go to problem 24.
If you answered **incorrectly**, review "Solving Quadratic-Like Equations" on page 63.

24. Solve for the values that satisfy the inequality $x^2 - x - 12 \le 0$.

Answer: $-3 \le x \le 4$

If you answered **correctly**, go to problem 26.
If you answered **incorrectly**, go to problem 25.

25. Solve for the values that satisfy the inequality $\dfrac{y}{y+4} \ge 0$.

Answer: $y < -4$ or $y \ge 0$

If you answered **correctly**, go to problem 26.
If you answered **incorrectly**, review "Quadratic and Other Inequalities" on page 66.

26. Solve $\sqrt{9x + 1} = x + 1$.

Answer: $x = 0$, $x = 7$

If you answered **correctly**, go to problem 27.
If you answered **incorrectly**, review "Radical Equations with Quadratics" on page 70.

27. Determine the domain and range of the function $f(x) = \dfrac{3}{x^2 - 4}$.

Answer: The domain, the x values, is all real numbers except 2 or –2. The range, the $f(x)$ values, is all real numbers except 0.

If you answered **correctly**, go to problem 29.
If you answered **incorrectly**, go to problem 28.

28. Given the function $f(x) = 2x^3 - x^2 + 3$, then $f(-2) = ?$

Answer: -17

If you answered **correctly**, go to problem 29.
If you answered **incorrectly**, review "Functions and Function Notation" on page 74.

29. Given the functions $f(x) = x^2 - 1$ and $g(x) = x^3 + 3$, then $(f + g)(x) = ?$

Answer: $x^3 + x^2 + 2$

If you answered **correctly**, go to problem 31.
If you answered **incorrectly**, go to problem 30.

30. Given the functions $f(x) = x^2 - 1$ and $g(x) = x^3 + 3$, then $(f - g)(x) = ?$

Answer: $-x^3 + x^2 - 4$

If you answered **correctly**, go to problem 31.
If you answered **incorrectly**, review "Function Operations" on page 77.

31. Given the functions $f(x) = x^2 - 1$ and $g(x) = x^3 + 3$, then $(f \circ g)(x) = ?$

Answer: $x^6 + 6x^3 + 8$

If you answered **correctly**, go to problem 33.
If you answered **incorrectly**, go to problem 32.

32. Given the functions $f(x) = 3x^2 - 3x$ and $g(x) = x + 2$, then $(f \circ g)(x) = $?

    Answer: $3x^2 + 9x + 6$

If you answered **correctly**, go to problem 33.
If you answered **incorrectly**, review "Composition of Functions" on page 79.

33. Given the function $f(x) = 3x^2 + x - 4$, find $\dfrac{f(x+h) - f(x)}{h}$.

    Answer: $6x + 3h + 1$

If you answered **correctly**, go to problem 34.
If you answered **incorrectly**, review "Difference Quotient" on page 80.

34. Find the inverse function, $f^{-1}(x)$ for $f(x) = \dfrac{x}{2-x}$.

    Answer: $f^{-1}(x) = \dfrac{2x}{x+1}$

If you answered **correctly**, go to problem 36.
If you answered **incorrectly**, go to problem 35.

35. Find the inverse function, $f^{-1}(x)$ for $f(x) = 5 + \sqrt[3]{x-3}$.

    Answer: $f^{-1}(x) = 3 + (x-5)^3$

If you answered **correctly**, go to problem 36.
If you answered **incorrectly**, review "Inverse Functions" on page 82.

36. Given the function $y = |x|$, write the new absolute value function that is down 4 units and left 2 units from the original.

    Answer: $y = |x + 2| - 4$

If you answered **correctly**, go to problem 38.
If you answered **incorrectly**, go to problem 37.

37. Given the function $y = \sqrt{x}$, write the new radical function that is right 1 unit and reflected over a vertical line from the original.

    Answer: $y = \sqrt{-(x-1)}$

If you answered **correctly**, go to problem 38.
If you answered **incorrectly**, review "Function Transformations" on page 85.

38. Use the Remainder Theorem to evaluate $f(-3)$ when $f(x) = x^6 + 4x^5 - x^4 - 11x^3 + 3x^2 + 2x + 6$.

    Answer: 0

If you answered **correctly**, go to problem 40.
If you answered **incorrectly**, go to problem 39.

39. Use the Remainder Theorem to evaluate $f(4)$ when $f(x) = x^5 - 4x^4 + 6x^3 - 26x^2 + 18x - 39$.

    Answer: 1

If you answered **correctly**, go to problem 40.
If you answered **incorrectly**, review "Remainder Theorem" on page 90.

40. Find all the solutions of $x^3 + 12x^2 + 41x + 30 = 0$.

   Answer: $x = -1$, $x = -5$, $x = -6$

If you answered **correctly**, go to problem 42.
If you answered **incorrectly**, go to problem 41.

41. Find all the solutions of $x^5 - 61x^3 + 900x = 0$.

   Answer: $x = 0$, $x = 5$, $x = -5$, $x = 6$, $x = -6$

If you answered **correctly**, go to problem 42.
If you answered **incorrectly**, review "Rational Root Theorem" on page 92.

42. Find the smallest integer upper bound for the roots of $3x^3 - 4x^2 + 5x - 3 = 0$.

   Answer: 2

If you answered **correctly**, go to problem 44.
If you answered **incorrectly**, go to problem 43.

43. Find the largest integer lower bound for the roots of $x^4 - x^3 + 5x^2 - 3x - 1 = 0$.

   Answer: -1

If you answered **correctly**, go to problem 44.
If you answered **incorrectly**, review "Upper and Lower Bounds" on page 96.

44. The vertex of the parabola $y = 6 - 2x - x^2$ is at what point?

   Answer: $(-1, 7)$

If you answered **correctly**, go to problem 47.
If you answered **incorrectly**, go to problem 45.

45. The graph of $y = x(x - 3)^2(x + 4)$ is below the x-axis for what values of x?

   Answer: $-4 < x < 0$

If you answered **correctly**, go to problem 47.
If you answered **incorrectly**, go to problem 46.

46. The graph of $30x - 7y = 21$ has an x-intercept at what point?

   Answer: $\left(\frac{7}{10}, 0\right)$

If you answered **correctly**, go to problem 47.
If you answered **incorrectly**, review "Using a Graphing Calculator to Graph Lines and Polynomials" on page 98.

47. Given the function $y = 3^x$, what is the x-value that makes $y = \frac{1}{9}$?

   Answer: -2

If you answered **correctly**, go to problem 49.
If you answered **incorrectly**, go to problem 48.

48. Given the function $y = 2^{-x}$, what is the value of y when x is 3?

Answer: $\frac{1}{8}$

If you answered **correctly**, go to problem 49.
If you answered **incorrectly**, review "Exponential Functions" on page 109.

49. Simplify $e\left(e^{2x}\right)^{2x}$.

Answer: e^{1+4x^2}

If you answered **correctly**, go to problem 52.
If you answered **incorrectly**, go to problem 50.

50. Simplify $\dfrac{e^{3x-1}}{e^{2x+2}}$.

Answer: e^{x-3}

If you answered **correctly**, go to problem 52.
If you answered **incorrectly**, go to problem 51.

51. Simplify $(e^{2x-1})(e^{2x-1})$.

Answer: e^{4x-2}

If you answered **correctly**, go to problem 52.
If you answered **incorrectly**, review "Exponential Functions" on page 109.

52. What is the total amount of money accumulated after 10 years if your deposit of $10,000 is earning 4 percent interest compounded monthly?

Answer: $14,908.33

If you answered **correctly**, go to problem 55.
If you answered **incorrectly**, go to problem 53.

53. What is the total amount of money accumulated after 3 years if your deposit of $100 is earning 10 percent interest compounded continuously?

Answer: $134.99

If you answered **correctly**, go to problem 55.
If you answered **incorrectly**, go to problem 54.

54. How long does it take for your investment to double if it's earning 5 percent interest per year compounded continuously?

Answer: About 13.9 years

If you answered **correctly**, go to problem 55.
If you answered **incorrectly**, review "Compound Interest" on page 114.

55. $\log_8 \frac{1}{2} = ?$

Answer: $-\frac{1}{3}$

If you answered **correctly**, go to problem 58.
If you answered **incorrectly**, go to problem 56.

56. $\log 100 =$

Answer: 2

If you answered **correctly**, go to problem 58.
If you answered **incorrectly**, go to problem 57.

57. $\log_a \dfrac{1}{a^2} =$

Answer: -2

If you answered **correctly**, go to problem 58.
If you answered **incorrectly**, review "Logarithmic Functions" on page 117.

58. Use the Laws of Logarithms to simplify $\log_2 2\sqrt{x-1}$.

Answer: $1 + \dfrac{1}{2}\log_2(x-1)$

If you answered **correctly**, go to problem 60.
If you answered **incorrectly**, go to problem 59.

59. Use the Laws of Logarithms to simplify $\ln\dfrac{1}{ex^2}$.

Answer: $-1 - 2 \cdot \ln x$

If you answered **correctly**, go to problem 60.
If you answered **incorrectly**, review "Laws of Logarithms" on page 119.

60. Solve for the value(s) of x: $125^{x+1} = 25^{2x+1}$.

Answer: 1

If you answered **correctly**, go to problem 63.
If you answered **incorrectly**, go to problem 61.

61. Solve for the value(s) of x: $\log_4(x+1) - \log_4 x = \log_4 16$.

Answer: $\dfrac{1}{15}$

If you answered **correctly**, go to problem 63.
If you answered **incorrectly**, go to problem 62.

62. Solve for the value(s) of x: $\ln(2x-1) = 0$.

Answer: 1

If you answered **correctly**, go to problem 63.
If you answered **incorrectly**, review "Solving Exponential and Logarithmic Equations" on page 124.

63. What is the x-intercept of the graph of $y = (x-3)^3$?

Answer: (3,0)

If you answered **correctly**, go to problem 66.
If you answered **incorrectly**, go to problem 64.

64. What is the *y*-intercept of the graph of $y = (x - 2)(x + 3)(x - 1)^4$?

    Answer: (0,–6)

If you answered **correctly**, go to problem 66.
If you answered **incorrectly**, go to problem 65.

65. Does the graph of $y = 4 - 3x^2 - x^3 + x^5$ rise infinitely high or drop infinitely low as the values of *x* get very large?

    Answer: Rises infinitely high

If you answered **correctly**, go to problem 66.
If you answered **incorrectly**, review "Graphing Polynomials" on page 130.

66. What is the vertical asymptote of the graph of $y = \dfrac{x - 3}{x + 2}$?

    Answer: $x = -2$

If you answered **correctly**, go to problem 68.
If you answered **incorrectly**, go to problem 67.

67. What is the horizontal asymptote of the graph of $y = \dfrac{x^2 - 1}{2x^2 + 3x - 2}$?

    Answer: $y = \dfrac{1}{2}$

If you answered **correctly**, go to problem 68.
If you answered **incorrectly**, review "Graphing Rational Functions" on page 135.

68. The graph of $y = \sqrt[3]{x + 1}$ crosses the *y*-axis at what point?

    Answer: (0,1)

If you answered **correctly**, go to problem 70.
If you answered **incorrectly**, go to problem 69.

69. The graph of $y = |2x + 3|$ has a minimum value when *x* equals what?

    Answer: $-\dfrac{3}{2}$

If you answered **correctly**, go to problem 70.
If you answered **incorrectly**, review "Graphing Radical Functions" and "Graphing Absolute Value and Logarithmic Functions" on pages 139 and 144, respectively.

70. The graph of $y = \log_3 x$ crosses the *x*-axis when *x* equals what?

    Answer: 1

If you answered **correctly**, go to problem 72.
If you answered **incorrectly**, go to problem 71.

71. The graph of $y = e^{x - 1}$ crosses the *y*-axis when *x* equals what?

    Answer: 0

If you answered **correctly**, go to problem 72.
If you answered **incorrectly**, review "Graphing Absolute Value and Logarithmic Functions" on page 144.

72. Rewrite the fraction by rationalizing the denominator: $\dfrac{36 - x}{6 + \sqrt{x}}$.

Answer: $6 - \sqrt{x}$

If you answered **correctly**, go to problem 74.
If you answered **incorrectly**, go to problem 73.

73. Rewrite the fraction by rationalizing the denominator: $\dfrac{5}{3\sqrt{15}}$.

Answer: $\dfrac{\sqrt{15}}{9}$

If you answered **correctly**, go to problem 74.
If you answered **incorrectly**, review "Radical Equations and Conjugates" on page 157.

74. Solve $\sqrt{16z + 9} - \sqrt{9z - 8} = 4$.

Answer: $z = 1$ or $z = \dfrac{513}{49}$

If you answered **correctly**, go to problem 76.
If you answered **incorrectly**, go to problem 75.

75. Solve $\sqrt{5 - 4x} = x + 10$.

Answer: -5

If you answered **correctly**, go to problem 76.
If you answered **incorrectly**, review "Radical Equations—Squaring More Than Once" on page 158.

76. Rewrite the radical in the form of a complex number: $\sqrt{-4}$.

Answer: $2i$

If you answered **correctly**, go to problem 77.
If you answered **incorrectly**, review "Complex Numbers" on page 160.

77. Simplify and rewrite the answer in the form of a complex number: $\dfrac{-4 \pm \sqrt{16 - 4(7)}}{2}$.

Answer: $-2 \pm \sqrt{3}\, i$

If you answered **correctly**, go to problem 79.
If you answered **incorrectly**, go to problem 78.

78. Simplify and rewrite the answer in the form of a complex number: $\sqrt{3 - 9(7)}$.

Answer: $2\sqrt{15}\, i$

If you answered **correctly**, go to problem 79.
If you answered **incorrectly**, review "Complex Numbers" on page 160.

79. Find the sum: $(-8 + 5i) + (6 - 2i)$.

Answer: $-2 + 3i$

If you answered **correctly**, go to problem 81.
If you answered **incorrectly**, go to problem 80.

80. Find the difference: $(-8 + 5i) - (6 - 2i)$.

    Answer: $-14 + 7i$

If you answered **correctly**, go to problem 81.
If you answered **incorrectly**, review "Operations Involving Complex Numbers" on page 162.

81. Find the quotient: $\dfrac{-8 + 5i}{6 - 2i}$.

    Answer: $-\dfrac{29}{20} + \dfrac{7}{20}i$

If you answered **correctly**, go to problem 83.
If you answered **incorrectly**, go to problem 82.

82. Find the product: $(-8 + 5i)(6 - 2i)$.

    Answer: $-38 + 46i$

If you answered **correctly**, go to problem 83.
If you answered **incorrectly**, review "Operations Involving Complex Numbers" on page 162.

83. Solve for the complex solutions of $2x^2 - 3x + 8 = 0$.

    Answer: $\dfrac{3}{4} \pm \dfrac{\sqrt{55}}{4}i$

If you answered **correctly**, go to problem 85.
If you answered **incorrectly**, go to problem 84.

84. Solve for the complex solutions of $x^2 + 16 = 0$.

    Answer: $\pm 4i$

If you answered **correctly**, go to problem 85.
If you answered **incorrectly**, review "Quadratic Formula and Complex Numbers" on page 164.

85. Solve for the value(s) of x: $\dfrac{4x + 5}{x + 1} - \dfrac{8}{2x - 4} = \dfrac{3x^2 - 10x}{(x + 1)(2x - 4)}$.

    Answer: $x = -2$ or $x = \dfrac{14}{5}$

If you answered **correctly**, go to problem 86.
If you answered **incorrectly**, review "Rational Equations" on page 166.

86. The variable y varies directly as the cube of x. When x is 2, y is 24. What is y when x is 3?

    Answer: 81

If you answered **correctly**, go to problem 88.
If you answered **incorrectly**, go to problem 87.

87. The variable y varies inversely as the square of x. When x is 2, y is 25. What is x when y is 1?

    Answer: 10 or -10

If you answered **correctly**, go to problem 88.
If you answered **incorrectly**, review "Variation" on page 168.

88. If the radius of a circle is 4 inches, then what is its circumference?

    Answer: 8π inches

If you answered **correctly**, go to problem 90.
If you answered **incorrectly**, go to problem 89.

89. If the diameter of a circle is 1 centimeter, then what is its area?

    Answer: $\frac{\pi}{4}$ square centimeters

If you answered **correctly**, go to problem 90.
If you answered **incorrectly**, review "Circle" on page 171.

90. What are the coordinates of the center of the circle $(x - 7)^2 + (y + 2)^2 = 100$?

    Answer: $(7, -2)$

If you answered **correctly**, go to problem 92.
If you answered **incorrectly**, go to problem 91.

91. What is the radius of the circle $(x + 2)^2 + \left(y + \frac{3}{4}\right)^2 = \frac{25}{36}$?

    Answer: $\frac{5}{6}$

If you answered **correctly**, go to problem 92.
If you answered **incorrectly**, review "Circle" on page 171.

92. What is the standard form for the ellipse $4x^2 + y^2 - 56x - 12y + 132 = 0$?

    Answer: $\dfrac{(x - 7)^2}{25} + \dfrac{(y - 6)^2}{100} = 1$

If you answered **correctly**, go to problem 94.
If you answered **incorrectly**, go to problem 93.

93. What is the center and the lengths of the horizontal and vertical axes of the ellipse $\dfrac{(x + 13)^2}{16} + \dfrac{(y - 14)^2}{9} = 1$?

    Answer: Center is at $(-13, 14)$. Horizontal axis is 8 units; vertical axis is 6 units.

If you answered **correctly**, go to problem 94.
If you answered **incorrectly**, review "Ellipse" on page 174.

94. What is the standard form of an equation for a parabola that has its vertex at $(-1, 2)$ and opens upward?

    Answer: $y = a(x + 1)^2 + 2$ where a is any positive number.

If you answered **correctly**, go to problem 96.
If you answered **incorrectly**, go to problem 95.

95. What is the vertex of the parabola $y = -3(x - 2)^2 + 5$, and does it open upward or downward?

    Answer: Vertex is $(2, 5)$, and it opens downward.

If you answered **correctly**, go to problem 96.
If you answered **incorrectly**, review "Parabola" on page 177.

96. What is the standard form of the hyperbola $100x^2 - y^2 + 600x - 4y + 796 = 0$?

   Answer: $(x + 3)^2 - \dfrac{(y + 2)^2}{100} = 1$

If you answered **correctly**, go to problem 98.
If you answered **incorrectly**, go to problem 97.

97. What is the center of the hyperbola $\dfrac{(x + 6)^2}{49} - \dfrac{y^2}{64} = 1$?

   Answer: $(-6, 0)$

If you answered **correctly**, go to problem 98.
If you answered **incorrectly**, review "Hyperbola" on page 179.

98. Solve the system of equations using the addition method: $\begin{array}{l} x - 3y = 15 \\ 4x + 3y = 20 \end{array}$.

   Answer: $x = 7$, $y = -\dfrac{8}{3}$

If you answered **correctly**, go to problem 100.
If you answered **incorrectly**, go to problem 99.

99. Solve the system of equations using the addition method: $\begin{array}{l} x + 3y = 1 \\ 3x - y = 13 \end{array}$.

   Answer: $x = 4$, $y = -1$

If you answered **correctly**, go to problem 100.
If you answered **incorrectly**, review "Solving Linear Systems Using the Addition Method" on page 193.

100. Solve the system using the substitution method: $\begin{array}{l} x - 5y = 13 \\ 3x - 4y = 6 \end{array}$.

   Answer: $x = -2$, $y = -3$

If you answered **correctly**, go to problem 102.
If you answered **incorrectly**, go to problem 101.

101. Solve the system using the substitution method: $\begin{array}{l} 2x + y = 11 \\ 3x - 4y = 11 \end{array}$.

   Answer: $x = 5$, $y = 1$

If you answered **correctly**, go to problem 102.
If you answered **incorrectly**, review "Solving Linear Systems Using Substitution" on page 195.

102. Solve the system using Cramer's Rule: $\begin{array}{l} 4x - 3y = 13 \\ 5x - 2y = 11 \end{array}$.

   Answer: $x = 1$, $y = -3$

If you answered **correctly**, go to problem 103.
If you answered **incorrectly**, review "Solving Linear Systems with Cramer's Rule" on page 196.

103. Solve the system of equations: $\begin{array}{l} 2x^2 + 4y^2 = 3 \\ 3x^2 - 8y^2 = 1 \end{array}$.

   Answer: $x = 1$, $y = \dfrac{1}{2}$ or $x = 1$, $y = -\dfrac{1}{2}$ or $x = -1$, $y = \dfrac{1}{2}$ or $x = -1$, $y = -\dfrac{1}{2}$

If you answered **correctly**, go to problem 105.
If you answered **incorrectly**, go to problem 104.

104. Solve the system of equations: $\begin{array}{l} x = y^2 + 2y + 3 \\ x = 4y + 2 \end{array}$.

Answer: $x = 6$, $y = 1$

If you answered **correctly**, go to problem 105.
If you answered **incorrectly**, review "Systems of Non-Linear Equations" on page 199.

105. The area of a rectangle is 9 more than its perimeter. The length is 3 greater than 4 times the width. What is the area of the rectangle?

Answer: The area is 45 square units. The length is 15, and the width is 3.

If you answered **correctly**, go to problem 106.
If you answered **incorrectly**, review "Story Problems Using Systems of Equations" on page 201.

106. Which of the following points are part of the solution of the system of inequalities $\begin{array}{l} x^2 + 2y^2 \geq 4 \\ 3x + y < 11 \end{array}$? Choices are A$(-1,1)$, B$(0,2)$, C$(1,1)$, D$(5,-4)$, E$(1,0)$, F$(-2,3)$, G$(-1,-2)$

Answer: Points B, F, G

If you answered **correctly**, go to problem 107.
If you answered **incorrectly**, review "Systems of Inequalities" on page 204.

107. Use the addition method to solve the system of equations: $\begin{array}{l} 3x + 2y - z = 11 \\ 4x - 3y - 7z = 9 \\ 2x + 5y + 6z = 5 \end{array}$.

Answer: $x = 1$, $y = 3$, $z = -2$

If you answered **correctly**, go to problem 109.
If you answered **incorrectly**, go to problem 108.

108. Use the addition method to solve the system of equations: $\begin{array}{l} 2a - 3b + c = 17 \\ 3a + 2c - 3d = 20 \\ a + 4b + 4d = -7 \\ 2b - 3c + d = -8 \end{array}$.

Answer: $a = 5$, $b = -2$, $c = 1$, $d = -1$

If you answered **correctly**, go to problem 109.
If you answered **incorrectly**, review "Systems of Equations with Three or More Variables" on page 209.

109. Use matrices to solve the system of equations: $\begin{array}{l} 4x + 2y - 3z = -23 \\ 5x - 3y + 4z = 11 \\ 3x + 5y + 2z = -15 \end{array}$.

Answer: $x = -2$, $y = -3$, $z = 3$

If you answered **correctly**, go to problem 110.
If you answered **incorrectly**, review "Solving Linear Systems Using Matrices" on page 217.

110. Given the matrices A $= \begin{bmatrix} 3 & -2 \\ -1 & 4 \end{bmatrix}$ and B $= \begin{bmatrix} 3 & 6 \\ -4 & 5 \end{bmatrix}$ find 2A $-$ 3B.

Answer: $\begin{bmatrix} -3 & -22 \\ 10 & -7 \end{bmatrix}$

If you answered **correctly**, go to problem 112.
If you answered **incorrectly**, go to problem 111.

111. Given the matrices $C = \begin{bmatrix} 3 & 4 & -5 & 1 \\ 0 & -2 & -6 & 8 \end{bmatrix}$ and $D = \begin{bmatrix} 3 & -3 & 2 & 0 \\ 0 & 4 & -1 & -1 \end{bmatrix}$ find $C + 4D$.

Answer: $\begin{bmatrix} 15 & -8 & 3 & 1 \\ 0 & 14 & -10 & 4 \end{bmatrix}$

If you answered **correctly**, go to problem 112.
If you answered **incorrectly**, review "Operations on Matrices" on page 220.

112. Find the inverse matrix A^{-1} for the matrix $A = \begin{bmatrix} 3 & -2 \\ -1 & 4 \end{bmatrix}$.

Answer: $\begin{bmatrix} \frac{2}{5} & \frac{1}{5} \\ \frac{1}{10} & \frac{3}{10} \end{bmatrix}$

If you answered **correctly**, go to problem 114.
If you answered **incorrectly**, go to problem 113.

113. Find the inverse matrix E^{-1} for the matrix $E = \begin{bmatrix} -5 & 2 \\ 7 & -3 \end{bmatrix}$.

Answer: $\begin{bmatrix} -3 & -2 \\ -7 & -5 \end{bmatrix}$

If you answered **correctly**, go to problem 114.
If you answered **incorrectly**, review "Inverse of a 2×2 Matrix" on page 223.

114. Given the matrices $A = \begin{bmatrix} 3 & -2 \\ -1 & 4 \end{bmatrix}$ and $C = \begin{bmatrix} 3 & 4 & -5 & 1 \\ 0 & -2 & -6 & 8 \end{bmatrix}$, find the product AC.

Answer: $\begin{bmatrix} 9 & 16 & -3 & -13 \\ -3 & -12 & -19 & 31 \end{bmatrix}$

If you answered **correctly**, go to problem 116.
If you answered **incorrectly**, go to problem 115.

115. Given the matrices $A = \begin{bmatrix} 3 & -2 \\ -1 & 4 \end{bmatrix}$ and $B = \begin{bmatrix} 3 & 6 \\ -4 & 5 \end{bmatrix}$, then find the product AB.

Answer: $\begin{bmatrix} 17 & 8 \\ -19 & 14 \end{bmatrix}$

If you answered **correctly**, go to problem 116.
If you answered **incorrectly**, review "Matrix Multiplication" on page 222.

116. Six friends just finished buying their textbooks for the new semester. The first friend bought the algebra, business, computer, and dietary books and also bought 6 packs of paper; she spent $327. The second friend bought the business, computer, dietary, and English books and also bought 4 packs of paper for a total of $268. The third friend bought the algebra, business, dietary, and English books plus 5 packs of paper; this cost him $320. The fourth friend bought the algebra, computer, and English books and 6 packs of paper for a total of $267. The fifth friend bought the business, computer, and dietary books plus 8 packs of paper; the total was $211. The sixth friend spent $202 on the algebra, dietary, and English books plus 1 pack of paper. How much did the different books and the paper cost?

Answer: Algebra $120, business $110, computer $70, dietary $15, English $65, paper $2 per pack

If you answered **correctly**, you are finished! Congratulations!
If you answered **incorrectly**, review "Applications of Systems of Equations" on page 225.

Index

Houghton Mifflin Harcourt
End-User License Agreement

READ THIS. You should carefully read these terms and conditions before opening the software packet(s) included with this book "Book". This is a license agreement "Agreement" between you and Houghton Mifflin Harcourt "HMH". By opening the accompanying software packet(s), you acknowledge that you have read and accept the following terms and conditions. If you do not agree and do not want to be bound by such terms and conditions, promptly return the Book and the unopened software packet(s) to the place you obtained them for a full refund.

1. **License Grant.** HMH grants to you (either an individual or entity) a nonexclusive license to use one copy of the enclosed software program(s) (collectively, the "Software") solely for your own personal or business purposes on a single computer (whether a standard computer or a workstation component of a multi-user network). The Software is in use on a computer when it is loaded into temporary memory (RAM) or installed into permanent memory (hard disk, CD-ROM, or other storage device). HMH reserves all rights not expressly granted herein.

2. **Ownership.** HMH is the owner of all right, title, and interest, including copyright, in and to the compilation of the Software recorded on the physical packet included with this Book "Software Media". Copyright to the individual programs recorded on the Software Media is owned by the author or other authorized copyright owner of each program. Ownership of the Software and all proprietary rights relating thereto remain with HMH and its licensers.

3. **Restrictions on Use and Transfer.**

   (a) You may only (i) make one copy of the Software for backup or archival purposes, or (ii) transfer the Software to a single hard disk, provided that you keep the original for backup or archival purposes. You may not (i) rent or lease the Software, (ii) copy or reproduce the Software through a LAN or other network system or through any computer subscriber system or bulletin-board system, or (iii) modify, adapt, or create derivative works based on the Software.

   (b) You may not reverse engineer, decompile, or disassemble the Software. You may transfer the Software and user documentation on a permanent basis, provided that the transferee agrees to accept the terms and conditions of this Agreement and you retain no copies. If the Software is an update or has been updated, any transfer must include the most recent update and all prior versions.

4. **Restrictions on Use of Individual Programs.** You must follow the individual requirements and restrictions detailed for each individual program on the Software Media. These limitations are also contained in the individual license agreements recorded on the Software Media. These limitations may include a requirement that after using the program for a specified period of time, the user must pay a registration fee or discontinue use. By opening the Software packet(s), you agree to abide by the licenses and restrictions for these individual programs that are detailed on the Software Media. None of the material on this Software Media or listed in this Book may ever be redistributed, in original or modified form, for commercial purposes.

5. **Limited Warranty.**

   (a) HMH warrants that the Software and Software Media are free from defects in materials and workmanship under normal use for a period of sixty (60) days from the date of purchase of this Book. If HMH receives notification within the warranty period of defects in materials or workmanship, HMH will replace the defective Software Media.

   (b) HMH AND THE AUTHOR(S) OF THE BOOK DISCLAIM ALL OTHER WARRANTIES, EXPRESS OR IMPLIED, INCLUDING WITHOUT LIMITATION IMPLIED WARRANTIES OF MERCHANTABILITY AND FITNESS FOR A PARTICULAR PURPOSE, WITH RESPECT TO THE SOFTWARE, THE PROGRAMS, THE SOURCE CODE CONTAINED THEREIN, AND/OR THE TECHNIQUES DESCRIBED IN THIS BOOK. HMH DOES NOT WARRANT THAT THE FUNCTIONS CONTAINED IN THE SOFTWARE WILL MEET YOUR REQUIREMENTS OR THAT THE OPERATION OF THE SOFTWARE WILL BE ERROR FREE.

   (c) This limited warranty gives you specific legal rights, and you may have other rights that vary from jurisdiction to jurisdiction.

6. **Remedies.**

   (a) HMH's entire liability and your exclusive remedy for defects in materials and workmanship shall be limited to replacement of the Software Media, which may be returned to HMH with a copy of your receipt at the following address: Software Media Fulfillment Department, Attn.: *CliffsNotes®* *Algebra II Practice Pack*, HMH Trade Reference, 9205 Southpark Center Loop, Orlando, FL 32819 or call 1-800-225-3362. Please allow four to six weeks for delivery. This Limited Warranty is void if failure of the Software Media has resulted from accident, abuse, or misapplication. Any replacement Software Media will be warranted for the remainder of the original warranty period or thirty (30) days, whichever is longer.

   (b) In no event shall HMH or the author be liable for any damages whatsoever (including without limitation damages for loss of business profits, business interruption, loss of business information, or any other pecuniary loss) arising from the use of or inability to use the Book or the Software, even if HMH has been advised of the possibility of such damages.

   (c) Because some jurisdictions do not allow the exclusion or limitation of liability for consequential or incidental damages, the above limitation or exclusion may not apply to you.

7. **U.S. Government Restricted Rights.** Use, duplication, or disclosure of the Software for or on behalf of the United States of America, its agencies and/or instrumentalities "U.S. Government" is subject to restrictions as stated in paragraph (c)(1)(ii) of the Rights in Technical Data and Computer Software clause of DFARS 252.227-7013, or subparagraphs (c) (1) and (2) of the Commercial Computer Software - Restricted Rights clause at FAR 52.227-19, and in similar clauses in the NASA FAR supplement, as applicable.

8. **General.** This Agreement constitutes the entire understanding of the parties and revokes and supersedes all prior agreements, oral or written, between them and may not be modified or amended except in a writing signed by both parties hereto that specifically refers to this Agreement. This Agreement shall take precedence over any other documents that may be in conflict herewith. If any one or more provisions contained in this Agreement are held by any court or tribunal to be invalid, illegal, or otherwise unenforceable, each and every other provision shall remain in full force and effect.

Practice *does* make perfect with CliffsNotes®!

CliffsNotes Practice Packs help you master all of your key subjects with hundreds of practice problems and their solutions.

Available wherever books are sold or visit us at CliffsNotes.com®

CliffsNotes